Neutrino

Neutrino

FRANK CLOSE

OXFORD
UNIVERSITY PRESS

OXFORD
UNIVERSITY PRESS

Great Clarendon Street, Oxford ox2 6DP
Oxford University Press is a department of the University
of Oxford. It furthers the University's objective of excellence in
research, scholarship, and education by publishing worldwide in

Oxford New York

Auckland Cape Town Dar es Salaam Hong Kong Karachi
Kuala Lumpur Madrid Melbourne Mexico City Nairobi
New Delhi Shanghai Taipei Toronto

With offices in

Argentina Austria Brazil Chile Czech Republic France Greece
Guatemala Hungary Italy Japan Poland Portugal Singapore
South Korea Switzerland Thailand Turkey Ukraine Vietnam

Oxford is a registered trade mark of Oxford University Press
in the UK and in certain other countries

Published in the United States
by Oxford University Press Inc., New York

British Library Cataloguing in Publication Data
Data available

Library of Congress Cataloging in Publication Data
Library of Congress Control Number: 2010930302

Typeset by SPI Publisher Services, Pondicherry, India
Printed in Great Britain
on acid-free paper by
Clays Ltd, St Ives plc

ISBN 978-0-19-957459-9

1 3 5 7 9 10 8 6 4 2

Contents

Ray Davis

With X-rays, which penetrate much more than ordinary light,
you can see inside your hand. With neutrinos, which penetrate
much more even than X-rays, you can look inside the Sun.

(Nobel Ceremony 2002)

Ray Davis was the first person to look into the heart of a star. He
did so by capturing neutrinos, ghostly particles that are produced
in the centre of the Sun and stream out across space. As you read
this, billions of them are hurtling, unseen, through your eyeballs
at almost the speed of light.

Neutrinos are as near to nothing as anything we know, and
are so elusive that they are almost invisible. When Davis began
looking for solar neutrinos, in 1960, many thought that he was
attempting the impossible. It nearly turned out to be so: 40 years
were to pass before he was proved right, leading to his Nobel
Prize for physics in 2002, aged 87.

Longevity is an asset in the neutrino business. Not everyone
would be so fortunate.

Foreword

In June 2006, I was invited by *The Guardian* newspaper to write the obituary of Ray Davis. I was surprised and honoured when, the following year, the obituary won the prize for the 'Best science writing in a non-scientific context'. A reason, I am sure, is that the story of Davis's remarkable career, in a sense, wrote itself.

An obituary necessarily focuses on the one person, but the saga of the solar neutrinos touched the lives of several others, scientists who devoted their entire careers to chasing this elusive quarry, only to miss out on a Nobel Prize by virtue of irony, chance or, more tragically, by having already died. For this quest spans half a century, Davis winning his Nobel at age 87. Of them all, the most tragic perhaps is the genius Bruno Pontecorvo. Although when I began writing *Neutrino* I anticipated that it would be the story of Ray Davis, I discovered that Pontecorvo seemed to be there behind the scenes to such an extent that this became his story too – and also the story of John Bahcall, Davis's lifelong collaborator who, to the surprise of many, was not included in the Nobel award. So I humbly dedicate this book to the memory of these three great scientists, whose own lives were testimony to what science is all about, and proof of Thomas Edison's assertion that genius is '1 per cent inspiration and 99 per cent perspiration'.

I am particularly indebted to four of my colleagues – whose own careers have been focused on neutrinos – for providing some of their own memories, and for correcting some of my misconceptions. If I have not succeeded, the fault is mine, not theirs: Nick Jelley, Peter Litchfield, Don Perkins and Jack Steinberger.

Oxford,   Frank Close
October 2009

1

A DESPERATE REMEDY

Of all the things that make the universe, the commonest and weirdest are neutrinos. Able to travel through the earth like a bullet through a bank of fog, they are so shy that half a century after their discovery we still know less about them than all the other varieties of matter that have ever been seen.

Some of these will-o'-the-wisps are coming up from the ground beneath our feet, emitted by natural radioactivity in rocks, and some are the result of radioactivity in our own bodies, but most of those hereabouts were born in the heart of the Sun, less than 10 minutes ago. In just a few seconds, the Sun has emitted more neutrinos than there are grains of sand in the deserts and beaches of the world, greater even than the number of atoms in all the humans that have ever lived. They are harmless: life has evolved within this storm of neutrinos.

Neutrinos can pass through the Sun almost as easily as through the earth. Within a few seconds of being born in the heart of the

Sun, these hordes have escaped from the surface and poured into space. If we could see with neutrino eyes, night would be as bright as day: neutrinos from the Sun shine down on our heads by day and up through our beds by night, undimmed.

Not just the Sun but each of the stars visible to the naked eye, and the countless ones seen by the most powerful telescopes, are all filling the void with neutrinos. Out in space, away from the Sun and stars, the universe is flooded by them.

Even you are producing them. Traces of radioactivity from potassium and calcium in your bones and teeth produce neutrinos. So, as you read this, you are irradiating the universe.

All in all, there are more neutrinos than any other particle we know, certainly far more than the electrons and protons that make the stars and all visible matter such as you and me. Once, they were thought to have no mass and to travel at the speed of light; today we know that they do have a little mass, though so trifling that no one has yet measured it. All we know is that if you had some subatomic scales, it would take at least 100,000 neutrinos to balance a single electron. Even so, their vast numbers make it possible that, in total, they outweigh all the visible matter of the universe.

The neutrinos from the Sun that have poured through you since you started reading this are already speeding onwards beyond Mars. A few hours from now they will cross the distant boundaries of the Solar System and head out into the boundless cosmos. If you were a neutrino, the chances are that you would be immortal, never bumping into atoms in billions of years.

Were you to ask a neutrino in the depths of space about its history, it is likely that it would turn out to be as old as the universe. The neutrinos born in the Sun and stars, numerous though they are, are relative newcomers. Most are fossil relics of the Big Bang, and have been travelling through space unseen for over 13 billion years. Neutrinos are passing through our universe like mere spectators, as if we were not here. They are so shy that it is remarkable that we know that they exist at all. How did these ghostly, invisible pieces of nothingness give themselves away? Why does nature need them? What use are they?

Nature hides its secrets deep, but there are clues; it's a matter of being prepared to notice and act on them. Five billion years ago, as the cocktail of elements from a supernova solidified into the rocks of the newborn Earth, radioactive atoms were trapped there. Radioactivity occurs when the nuclei of atoms spontaneously change form: granite is not forever the same. For as long as the earth has existed, atoms of uranium and thorium, frozen into the minerals of its crust, have been eroding, transmuting into lighter elements, cascading down the periodic table until they have changed into stable atoms of lead. It is in this natural chronometer of radioactivity that neutrinos are born. That is where our story begins.

Radioactivity

Chance plays a leading role in science, but to gain the glittering prizes it is not sufficient to be in the right place at the right time; you must also be able to recognise the gifts that serendipity presents. Had Röntgen not glanced out of the corner of his eye as he closed the door of his dark laboratory in November 1895, or not given further thought to the faint glimmer that had momentarily captured his attention, he would not have discovered X-rays. Röntgen had found that when a flow of electrons hits glass, it could produce mysterious rays capable of penetrating solid matter, such as skin. This bizarre phenomenon, able to display broken bones as shadows on photographic emulsion, started modern atomic science and inspired the work that led to the discovery of radioactivity.

Here too chance entered. The news of X-rays was sensational, and they were the centre of attention when the French Academy of Sciences met on 20 January 1896. At that meeting was Henri Becquerel who had followed his father's interest in phosphorescence – the ability of some substances to glow after exposure to light, in effect to store up radiation. No one had any clear idea about what X-rays were, but there was a lot of discussion as to whether they were associated with the glass in Röntgen's apparatus phosphorescing. Becquerel immediately realised that here was a puzzle made for him. He had some phosphorescent crystals that he had prepared with his father years before, and so he set out to see if any of them emitted X-rays. The specimen was a compound containing potassium, sulfur and uranium.

That was his first piece of good fortune. The element uranium would turn out to be crucial.

He put the phosphorescent substance on top of a photographic plate that had been wrapped in paper to protect it from the light, and left them in the Sun. The sunlight energised the phosphorescent material but not the plates, so when he developed them he was excited to find a smudgy image. When he placed a piece of metal between the material and the plate, a clear outline of that too could be seen. His immediate reaction was that sunlight had stimulated the emission of X-rays, which had penetrated the paper but not the metal – hence the shadow.

It was at this point that more luck enters the story. Typical winter weather set in and, during the last part of February, Paris was overcast for several days. Without sunlight, Becquerel could not energise his specimen; it would be impossible to induce phosphorescence and therefore the X-rays – or so he thought. He had kept the sample in a cupboard hoping for a bright day, but none came. Eventually he gave up and on 1 March, tired of waiting, decided to develop the plate anyway. Becquerel's son recorded that Henri was 'stupefied' to find the pictures of silhouettes were even more intense than he had obtained earlier that month in the sunshine.[i]

Whatever the radiation was, it had no need for sunlight. It appeared spontaneously, without any prior stimulation. This was utterly novel. Röntgen's X-rays were the result of an electric current having first supplied energy to glass; phosphorescence was the result of sunlight giving energy to materials; Becquerel's radiation appeared to come for free.

Becquerel had had two pieces of fortune: he was using uranium, which emits radiation without prior stimulation, and the dark days had metaphorically brought this to light. A third piece of good fortune was to avoid the mistake of assuming that the fogging was due to poor quality plates. This was of course possible, even likely, and so the use of the piece of metal was crucial; its shadow showed that there were genuine rays coming from above, and that the image on the photograph was not some inherent blemish. This at least was not luck, but an example of careful science, as a result of which Henri Becquerel discovered radioactivity.

However, he did not give it this name (that would come later from Marie and Pierre Curie), nor did he have a clue what it was. Indeed, most people ignored him. During the previous years several weird phenomena had shown up, such as fluorescence and X-rays, so a new type of radiation did not seem particularly special. This one, however, was to prove momentous.

Alpha Beta Gamma

In many detective stories, the supposed perfect crime has been solved by pursuing some trifling clue left at the scene. Becquerel had found a mere smudge on a photographic plate, so modest that it could easily have been overlooked, yet in this trifling whiff of radiation, it would turn out that Nature had exposed the route to the secrets of creation. Of course, neither Becquerel nor anyone else knew or even suspected that at the time. All he had was a cloudy image, and the immediate challenge was to understand what it meant.

Marie and Pierre Curie chased the source of the radiation by separating elements in pitchblende – a radioactive substance – finding which samples were more radioactive, and then selectively refining them until the concentration of radiation grew. As a result Marie found a new element, polonium, which was highly radioactive. Even better, she found radium. If there had been any controversy about the reality of radioactivity before, all doubt disappeared with the discovery of radium. Radium is so radioactive that, when held in the hand, it feels warm. This heat shows that radioactivity releases energy from the substance spontaneously, day in day out. Marie Curie was naively unaware of the implications of this power; years were to pass before the effects of that radiation on the body would be realised, by which time it was too late: she was already showing signs of radiation sickness.

The discovery of radium had two important consequences. First, it showed that radioactivity, as the Curies named it, is not restricted to uranium; it is a property of nature whereby some elements can spontaneously emit energy without prior stimulation.

Second, no longer was science restricted to smudges on a photographic plate; the radioactivity of radium was so powerful that its effects could be felt, measured and analysed. Now science could advance in its forensic way. The person who identified the nature of radiation and exploited it almost single-handedly was Ernest Rutherford. As a student in his native New Zealand, in 1895 he had discovered how to detect radio waves, many years ahead of Marconi.[1] Rutherford came second in the competition for the scholarship named after the 1851 Exhibition, which enabled new graduates to continue their studies abroad. Fortunately for him, in what turned out to be a seminal moment in the history of science, the winner that year, J C Maclaurin, decided to stay in New Zealand for family reasons. So Rutherford took up the award and duly arrived in Cambridge in September 1895 intending to work on radio. Those were his plans, but Röntgen had just discovered X-rays, and Becquerel soon followed with his discovery of radioactivity. J J Thomson, head of the group and himself about to discover the electron, suggested that Rutherford should work on these new radiations. This was settled once Lord Kelvin, the leading scientist of the age, famously opined that there was 'no future in radio'.

So Rutherford set to work unravelling the inner labyrinths of the atom, leaving Marconi to prove Lord Kelvin wrong. Had

[1] As an aside, I should mention Oliver Lodge. He produced and detected electromagnetic waves in 1888, before Hertz, but instead of publicising the fact he went on vacation. Hertz published first, and his name is forever associated with them. In 1894, at the meeting of the British Association in Oxford, Lodge demonstrated the transmission of signals – little more than Morse code – over a distance of 50 metres. He later admitted that he had not seen any useful application of the phenomenon, and so failed to realise the potential of wireless communication.

Rutherford replaced Marconi in the history of radio, others would presumably have their names attached to the sequence of discoveries on the nature of radioactivity, the nuclear atom, transmutation of the elements and the power within the atom, all of which are associated with Rutherford. His first contribution to this new science was in showing that radioactivity held more surprises than anyone expected. For a start, it came in three different forms.

A thin sheet of paper is enough to cut off some of the radiation almost immediately. I say 'some' because there remained a more penetrating radiation which only died off gradually. Rutherford revealed the forms with startling simplicity, covering the uranium with thin sheets of aluminium foil, and gradually increasing their number. For the first three foils he found that the strength of the radiation died away progressively: the thicker the layer of aluminium, the less radiation penetrated. However, as he added further layers, the radiation appeared to maintain its intensity, only gradually falling off after several more foils had been added. He realised that there must be 'at least two distinct types of radiation – one that is very readily absorbed which will be termed for convenience the alpha radiation, and the other of a more penetrating character which will be termed the beta radiation'. He later discovered a third form, which he duly named gamma.

Today, we know that these three forms of radiation are caused by three different forces. These are respectively the strong, weak and electromagnetic forces. Together with gravity, these form the four fundamental forces of Nature, which build atoms and bulk matter, and control the workings of the universe. It is remarkable that Rutherford distinguished among these in his very first atomic experiments.

Naming things gives an illusion of understanding, but is merely classification. Nonetheless, it is an important first step, which inspires questions such as what gives the differing attributes associated with the different names? The differences eventually became literally visible when Charles Wilson put a radioactive source inside a 'cloud chamber'. In the supersaturated vapour of the chamber, electrically charged particles in motion leave ephemeral vapour trails. Wilson described them as 'little wisps and threads of cloud'. The alpha radiation left strong thick trails and the beta trails were thinner and wispy whereas the gamma rays left no trails but gave themselves away when they bumped into electrons in atoms and set these in motion. Magnetic fields would curve the paths, showing that the alpha and beta radiations respectively, consisted of positively and negatively charged particles, while the absence of trails for gamma rays is because they have no electrical charge. Rutherford exclaimed that 'at last we have a telescope to look inside the atom'.

The alpha particles turned out to be relatively massive and, we now know, are pieces of atomic nuclei. They consist of tight bundles of two protons and two neutrons emitted when the strong forces that hold an atomic nucleus together are disrupted. When this happens, the large nucleus of a heavy element can spontaneously change into a smaller slightly lighter one by ejecting the tight bundle – the alpha 'particle'. Being positively charged, the alpha particle can attract two negatively charged electrons and form an atom of helium. We now know that helium gas found in some rocks on Earth is the result of such nuclear transmutations.

Rutherford was to later gain fame for discovering the atomic nucleus, using alpha particles as probes of the atom.[ii] The beta radiation consists of electrons, not ones that pre-existed in the

Figure 1 Trails in a Cloud Chamber.

atom but which have been created[2] from energy released in the nuclear transmutation: alchemy. Gamma rays are particles of light, far beyond the rainbow, having much shorter wavelengths than visible light. So three varieties of radiation had been identified but no one suspected that the beta radioactivity also contained a ghost at the feast.

$E = mc^2$

Isaac Newton in the 17th century had realised the importance of energy. Push something and, in the absence of any friction, it

---

[2] Dmitrij Iwanenko first proposed the idea that electrons are created in beta-decay, just like photons are created in atomic transitions, and as such don't 'pre-exist' in an atom.

will start to move. Keep pushing and it will speed up. Newton defined energy of motion, kinetic energy, as proportional to the amount of force that you pushed with, and the distance over which you kept pushing the object. He was also aware that energy could have different manifestations. A body on top of a cliff has potential energy – the potential to gain kinetic energy if it falls over the edge. Potential energy is in proportion to height above some ground level: the higher you are the more potential energy you have. As you fall towards ground, the force of gravity accelerates you. You gain kinetic energy at the same rate that you lose potential energy; the sum is preserved. This is a simple example of energy conservation, and of the change from one form of energy to another, in this case from potential to kinetic.

There are many other ways that energy can be redistributed. In the 19th century, thermodynamics – the science of heat and motion – matured. Energy in the form of heat can be converted into kinetic energy. The steam engine works on this principle. When water boils, it turns into steam and expands. If the expansion is in a closed cylinder whose end can move, the pressure of the steam can force the piston into motion. Attach the moving end to a rod, which in turn is connected to a wheel, off-centre, and the result will be that the wheel turns. By this means, steam power enabled trains, weighing hundreds of tons, to travel at over 100 kilometres an hour.

In the steam engine, as in countless other examples, energy is being changed from one form to another, but overall it is conserved. That is the first law of thermodynamics on which whole industries have been built. It is one of the most fundamental and far-reaching laws of nature.

While all the excitement about radioactivity was happening, and independent of it, in 1905 Albert Einstein announced his

Theory of Special Relativity. Its most famous equation, $E = mc^2$, implies a profound link between energy and mass: that mass (m) and energy (E) can be converted one into the other at an exchange rate governed by the speed of light (c). Einstein's equation expressed a new and profound way of storing and transferring energy, but here again, energy overall is conserved.

Radioactivity is an example of $E = mc^2$ at work. When the matter in the nucleus of an atom spontaneously rearranges itself, the energy that had been, a moment earlier, locked within some of the original mass is suddenly released. It may be radiated as light – gamma rays; it may be taken up as kinetic energy as pieces of the previous nucleus shoot off, as in alpha decay; or it may congeal into new forms of matter, as in beta decay.

In alpha and gamma decay, the energy accounts were straightforward; in beta decay, however, they seemed not to work. If there is only the one particle emitted each time that a radioactive nucleus decays, energy conservation enforces a single value for its energy. That is what was seen in alpha and gamma decays, but in 1914, James Chadwick discovered that the energy of beta radiation varied from one measurement to the next. Instead of always having the same energy, electrons emerged with a continuous range of energies, sometimes almost no energy at all, and on other occasions amounts all the way up to a maximum value.

Neils Bohr, who earlier had fathered the model of the atom as electrons 'orbiting' Rutherford's central nucleus, put his authority behind a radical suggestion: energy is not conserved in beta decay.

This ran counter to centuries of experience, and was an act of desperation. The Austrian theorist, Wolfgang Pauli, refused to accept it, and put forward another explanation. He proposed that the beta particle was accompanied by an 'additional very penetrating radiation that consists of new neutral particles'. In such an eventuality, energy is conserved but is being shared between two particles rather than carried off entirely by just one. In Pauli's theory the visible particle, the beta, sometimes carried all of the available energy leaving nothing for the invisible neutral partner, while on other occasions the invisible one took away some of the energy leaving less for the beta particle. As a result the energy carried by the visible beta particle could be anywhere within a range, rather than being restricted to a single value.

This sounds like a conservative idea, and fitted the facts, but at the time it was greeted with little more enthusiasm than Bohr's proposal. The reason was that it ran counter to the prevailing beliefs about the nature of atoms. The rich tapestry of Nature at that time appeared to be made of just two particles: electrons and protons. This fundamental simplicity promised a beautiful unification at the core of matter, whereas introducing a third particle for no reason other than to fix up one esoteric puzzle, seemed to many to be unwarranted.

Pauli and the Neutrino

Pauli was born in Vienna in 1900. A remarkably clear thinker, at the age of 19 he wrote the best explanatory textbook on Special Relativity, which nearly a century later is still a classic. By 22, he had a PhD and was working on the foundations of the new quantum mechanics, later winning a Nobel Prize.

Pauli was also infamous for his acid comments about other scientists' work, once damning a concept that was so vague that it was untestable and hence of no use to science with the remark that it was 'not even wrong'. Ironicly such a criticism might have justifiably been levelled against his solution to the mystery of the disappearing energy in beta decay: having proposed an invisible particle, he even wagered a case of champagne that no one would ever be able to detect the beast.

After Rutherford's experiments had showed that atomic nuclei are made of constituent particles, the world view was that these consisted of protons and electrons. Rutherford himself thought so. The proton was the massive core at the heart of the simplest atom of hydrogen, but he realised that the masses of the nuclei of heavier elements could only be explained if there was also some neutral particle of similar mass to the proton. Rutherford named it the 'neutron'. His picture was that a neutron was some tightly bound combination of a single proton and an electron.

This idea fell apart in 1927. The electron and proton had each been found to spin, and always with the same rate. This was soon explained theoretically by the mathematician Paul Dirac as a consequence of quantum mechanics and relativity.[3] What also became obvious was that a neutron could not be a combination of these two. The reason had to do with what was known as the 'nitrogen anomaly'.

The rates at which various atomic nuclei spin had been measured and showed that a nucleus of nitrogen must contain an even number of spinning constituents. Chemistry showed that a nitrogen atom contains seven electrons, and so its nucleus must have seven protons to counterbalance the electric charge. If this had been the whole story, a nitrogen nucleus would only have

---

[3] See my book *Antimatter* for an explanation of this. For the life story of Dirac, see G Farmelo, *The Strangest Man*.

been half as massive as in reality. So seven neutrons were called for. If neutrons were single beasts, like protons, this $7 + 7 = 14$ would satisfy the even-number rule. However, if each neutron was really a pair, the total number of constituents would become 21, an odd number. Rutherford's picture of a proton–electron combination simply didn't fit the facts.

This is where Wolfgang Pauli enters the story, inventing a new neutral particle which, he initially thinks, can solve two puzzles for the price of one particle.

Pauli made his proposal in a letter of 4 December 1930, whose primary purpose was to apologise for being unable to attend a meeting on radioactivity in Tubingen because 'I am indispensable here in Zurich because of a ball on the night of 6/7 December'. Beyond the social aspects, which were most brief, the letter shows both the radical nature of his proposal and also the state of knowledge at the time. It also illustrates with hindsight how science can advance by a mixture of genius and confusion.

Pauli began by addressing the problem in nuclear physics, where the properties of nitrogen nuclei did not fit well with the idea that nuclei are made from just protons. (Indeed, this problem was not restricted to nitrogen; a similar anomaly arose with lithium.) Pauli realised that all would be well if Rutherford's model of a proton–electron combination were given up and the neutral object were regarded as a single particle, identical to a proton in all respects but for its electrical neutrality. He proposed that there 'exist in the nucleus electrically neutral particles, that I wish to call neutrons'. He described them as like protons but without electric charge, adding that they 'differ from light quanta in that they do not travel with the velocity of light'. The massive neutron, an electrically neutral partner to the proton, was soon discovered – in 1932, by James Chadwick, the same person who, in 1914, had discovered the anomalous energy behaviour in beta-decays. It is an essential constituent of all nuclei (save that of hydrogen, which usually consists of a single proton). What we call isotopes are nuclei with a given number of protons, whose number determines which element is seeded, but with different numbers of neutrons. Thus, uranium 235 and 238 each contain 92 protons, which is what makes them uranium, but have 143 and 146 neutrons respectively giving a total number of 235 or 238 constituents. The neutron is today recognised as a central player in nuclear physics.

So far, so good, for the nucleus. However, Pauli also proposed that this same neutron was produced along with the electron in beta decay. The modern neutron is identical to his first proposal – the neutron as constituent of the atomic nucleus – however, it is not the same as the mystery guest in beta decay, the particle that we now call the neutrino. In 1930, however, Pauli knew none of this. He referred to both players as neutrons, as in the following abst...ract[4] (where I have used his words but put [...] around 'neutron' when it refers to what eventually became known as a neutrino):

The continuous beta spectrum would then become understandable by the assumption that in beta decay a [neutron] is emitted in addition to the electron such that the sum of the energies of the [neutron] and the electron is constant...I agree that my remedy could seem incredible because one should have seen those [neutrons] already if they really exist. But only the one who dares can win...every solution to the issue must be discussed. Thus, dear radioactive people, look and judge.

[4] Pauli's letter and his personal impressions of the history are contained in his technical article in *Neutrino Physics*, edited by Klaus Winter, Cambridge University Press.

16

Hans Geiger, who had worked with Rutherford in discovering the atomic nucleus, was at the meeting. He realised that Pauli's solution to the energy accounts for beta decay might work, and wrote him a letter. Years later Pauli recalled his excitement at having received it, but at the time seems not to have appreciated the importance because no copy of Geiger's letter has survived.[5] Possibly his enthusiasm had been dampened by the fact that he had already realised that the neutral particles involved in the beta decay could not be the same as his hypothesised nuclear constituent, the neutron. The nuclear masses needed a neutral particle whose mass was equal, or at least very similar, to that of a proton, which is what Chadwick was about to discover. However, Pauli's explanation of beta decay required a neutral particle that had no mass at all, or at most a trifling amount.

Pauli continued to mention his idea, to see how other scientists responded. Few liked it, opinions ranging from 'simply wrong' to 'crazy'.[6] It was in October 1931, at a meeting in Rome where he talked with Enrico Fermi, that things began to fall into place.

Pauli later recorded that Fermi 'immediately expressed a lively interest in my idea'. Neils Bohr was less impressed. Inventing new particles to fix fundamental problems was not his style. He had seen the subtle way that the energy accounts could borrow and repay in atomic physics, and therefore saw no reason why energy conservation might not actually apply in the even stranger world of atomic nuclei. Fermi and Pauli discussed this together but did not like it. Bohr seemed happy enough to accept that

[5] W Pauli in *Neutrino Physics*, ed. K Winter, Cambridge Monographs on Particle Physics, Nuclear Physics and Cosmology, (1998) p. 14 where Pauli thanks 'Mrs Meitner for keeping a copy of this letter and leaving it to me'.

[6] G Farmelo *The Strangest Man*, p. 195.

electric charge was conserved in nuclear processes, so why not energy? Fermi felt that Pauli's idea made more sense.

When, in 1932, James Chadwick discovered that there is indeed a neutron in the nucleus, but that it is heavy, this was a mixed blessing. The good news was that Pauli was right – at least as regards the neutron in the nucleus. The downside was that it could not also be the lightweight particle that he wanted for explaining beta decay. However, the appearance of the neutron had increased the number of atomic particles by 50% and the idea of inventing a further particle no longer seemed so heretical.

Once Chadwick had discovered this genuine nuclear constituent, Pauli stopped using the name neutron for the particle that was his solution to the beta-decay puzzle. The beta decay does include a lightweight neutral particle, as Pauli suggested, but it does not pre-exist in the nucleus any more than a bark exists in a dog. Pauli duly dropped the name neutron for it, but he had no special alternative. Fermi however did. To differentiate Pauli's proposed lightweight neutral particle from the massive neutron he dubbed it the 'little neutron'; in Italian: neutrino.

SEEING THE INVISIBLE

In 1911, the Belgian industrialist Ernest Solvay invited about twenty of the world's leading physicists to a conference in Brussels. This was the first of the 'Solvay Conferences', which would become famous for their singular role in charting the course of science throughout the 20th century. In 1927 and 1930, the theme was quantum mechanics, which had just burst onto the scene, providing the long-sought equations that explain the behaviour of electrons in atoms. The intention was that the conference in 1933 would focus on the application of quantum mechanics to chemistry. However, a torrent of unexpected discoveries caused a last minute change of plan. Pauli had invented the neutrino in 1930. In 1932 the neutron and also the first example of antimatter – the positive analogue of the electron, known as the positron – were both found. Experiments with the first 'atom smasher' had shown that the atomic nucleus has a rich and complex structure, which could be altered by human action as well as by spontaneous radioactivity. In 1933, Irene Joliot-Curie, daughter of Marie Curie, and her husband Frédéric Joliot showed that, in such examples of 'artificial radioactivity', beta decays could produce the positively charged positron as easily as the familiar negatively charged electron. As a result, two dozen of the world's leading physicists met at the Solvay Conference during the week 22–29 October 1933 to discuss not quantum chemistry, but a new science: nuclear physics. The roll-call included Einstein, of course, Rutherford, the father of nuclear physics, and Marie Curie, sick from radiation and terminally ill. Also present were Pauli, Fermi and Bohr. It was following discussions among this latter trio during the congress that the idea of the neutrino began to mature into hard science. It was primarily Fermi who cleared the mists as a result of what he learned during that week. His inspiration began when Frédéric Joliot described his discovery that beta decays could occur in two distinct ways. The emission of negative rays, which consisted of the well-known electrons, was what had exercised everyone to date, but now Joliot showed how he had found examples where the new positron emerged. Apart from the appearance of a positive positron instead of a negative electron, everything else looked pretty much the same.

Fermi excelled in mental imagery, and Joliot inspired him to visualise nuclei made of protons and neutrons, which then changed their nature by beta decay. He realised that this implied a profound symmetry. If a neutron changed into a proton, the total electric charge would be balanced by emission of a negatively charged electron – the familiar beta particle; but why not also imagine that, in suitable circumstances, a proton in a nucleus could turn into a neutron? In this case, the charge would be balanced by a positively charged beta ray – the positron. For Fermi, the neutron, proton and beta particles – whether electron or positron, it was just the electric charge that mattered – were the central players in these nuclear processes.

That was just the first of his insights. Everything fell into place when some further news arrived.

Pauli had realised that there might be a way to tell if a lightweight neutrino was accompanying the emission of an electrically charged beta particle. If the energy spectrum of the beta rays could be measured very precisely, one might discover whether their energies continued all the way to a maximum, and then stopped, or instead carried on to infinite energy. Bohr believed that energy conservation was only true when averaged over large numbers of events, being violated on an event-by-event basis; consequently, the spectrum of the beta particle energies could extend onwards for ever.[7] However, if the energy spectrum ended sharply at some finite amount, Pauli would be vindicated. Pauli's suggestion that physicists should measure the energy spectrum very carefully at the high energy end to see if it went smoothly onwards or stopped suddenly, had been taken up. The results were announced to the conference: there was indeed a clear upper limit to the spectrum.

This was music to Pauli's ears, and convinced him that his idea of an unseen third guest at the party was correct. In the ensuing discussion he stood up and announced his idea of the neutrino:

their mass cannot be very much more than the electron mass. In order to distinguish them from heavy neutrons, Mr Fermi has proposed to name them 'neutrinos'. It is possible that the proper mass of neutrinos be zero...It seems to me plausible that neutrinos have a spin ½... We know nothing about the interaction of neutrinos with the other particles of matter or with photons.

[7] Pauli realised that Bohr's hypothesis would imply a 'Poisson' distribution, characteristic of statistical effects.

Everything was now in place for Fermi to make his theory of beta decay in which Pauli's neutrino would play a central role.

Fermi's Theory

Fermi started on this immediately after the Solvay conference. He took everything that Pauli had suggested, together with what he had learned at the congress – the role of neutron and proton, the conviction that there was indeed a neutrino in the beta decay, and also the emerging theory of quantum electrodynamics – to develop his idea. He assumed that energy and momentum are conserved in beta decay, and that rotation – angular momentum or spin – is also conserved.

Particles have an intrinsic angular momentum: spin. Quantum theory shows that this can only take on certain specific values that are either odd or even multiples of a basic unit. For historical reasons, this basic unit of spin is known as ½ and so, odd multiples are half-integers, while even multiples are integers. Today, particles that belong to the former class are known as fermions, after Enrico Fermi; those in the latter are bosons, named after the Indian theoretician, Satyendra Bose. It is the fermions that are the main players in our story.

The proton is a fermion with spin ½. It had been the anomaly with the spin of the nitrogen nucleus that had led Pauli to propose the neutron, which also has spin ½, in order to get the total spin of that nucleus correct. The electron too has spin ½, a fact deduced from atomic spectra and from the way that atoms respond to magnetic fields.

The rules of spin in quantum mechanics are that two halves can make a whole but you need three to combine to make a half.

So the beta decay of a neutron into a proton and an electron cannot be the whole story: the neutron at the start has spin ½ and so an odd number of spin ½ particles must emerge when it decays. The proton and electron therefore needed to be accompanied by a third particle, with spin ½ and no electric charge: the neutrino.

Fermi had identified the actors. Now he made the first attempt to work out the plot. His idea also built on the observation that a neutron appears like a proton with its electric charge removed, and he guessed that the neutrino is similarly related to the electron. He then used this parallelism between the electron and neutrino, and between the proton and neutron, together with the new and successful theory of electrically charged particles and light – quantum electrodynamics – as the basis of his theory of beta decay. He assumed that the four particles could momentarily occur at the same point in space and time. In this scheme, a neutron could spontaneously transmogrify into a proton, emitting an electron (the beta particle) and a neutrino (the ghost).

Today, we know that this is not the whole story, as there is a small gap between the place where the neutron turns into a proton, and the location where the liberated energy and electric charge rematerialise as an electron and neutrino. However, this gap is smaller than the size of a neutron, and in Fermi's day it

Figure 2 Fermi's Model of Beta Decay. In Fermi's model, a neutron denoted n^0 turns into a proton p^+, electron e^-, and neutrino v^0 at a single point in space. The superscripts denote the amount of electric charge that each particle has relative to that of a proton, and the sign denotes whether it is negative or positive.

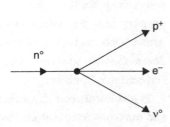

was not possible to resolve the size of an entire nucleus, let alone an individual neutron or proton. In fact, Fermi's model was so good that even today it remains the standard introduction to the theory of beta radioactivity for undergraduate physics students.

Starting with this theory, he was able to calculate what the energy spectrum of electrons produced in beta decays should look like. It turned out to be as the experiments had found, including the cut-off at the high energy end. Putting it all together implied that the mass of a neutrino could be at most a tiny proportion of that of the electron, and could possibly be nothing at all. Even more careful measurements were made, and when compared with Fermi's theory they showed that the neutrino spins at the same rate as the neutron, proton and electron. Everything that Fermi had assumed was turning out to be true.

Despite these successes, many physicists did not believe that the neutrino was real. A free neutrino being absorbed by something, affecting pre-existing matter by bumping into it and changing something, thereby revealing its own existence, was missing. The general lack of enthusiasm for the neutrino at that time was shown when Fermi produced his paper in 1934.

Entitled 'Tentative theory of beta rays', it was sent to the leading English-language scientific journal, *Nature*. The editor rejected one of Fermi's greatest pieces of theoretical physics, having received advice that the manuscript 'contained speculations too remote from reality to be of interest to the reader'. Half a century later the editors would admit this to have been their greatest blunder. The paper eventually appeared in Italian in *Nuovo Cimento*, and soon after in German in *Zeitschrift fur Physik*, but never in English.

What Fermi had done in his theory was to take the idea of the neutrino seriously and propose how the recently discovered neutron, and the laws of quantum mechanics, allowed a neutron in a nucleus to convert spontaneously into a proton, emitting an electron (the beta particle) and a neutrino. Speculative, certainly; untestable, as Pauli had speculated, possibly; but 'too remote from reality' and not of 'interest'? Certainly not.

This saga had so exhausted Fermi that he decided to switch from theory to experiments 'for a short while'.[8] As it turned out, the experiments became an all-consuming project that would keep him busy for years and eventually lead the German scientists, Otto Hahn and Fritz Strasseman, and the Austrian-born Lise Meitner, to discover uranium fission, with all that that would lead to. However, Fermi's theory was not forgotten and, as it turned out, had opened the way for Pauli's hypothesis of the neutrino to become scientifically tested.

The Neutrino Starts to become Real

Fermi's hypothesis that the four actors could meet and swap identities at a point did more than just describe beta decay: the theory implied that a neutrino could bump into a neutron and convert it into a proton and an electron. This is like beta decay in reverse. Suddenly, with Fermi's theory, the neutrino has ceased to be just

---

[8] As quoted in Laura Fermi, *Atoms in the Family*. Fermi had a remarkable gift for both experiment and theory. An example is what happened when the first atomic bomb was exploded in the desert of New Mexico. Fermi, along with some of the greatest scientists of the time, was hiding in a bunker miles from the blast. While everyone was astonished by what they saw, Robert Oppenheimer making his famous quotation from *The Gita*: 'now I am become death, the destroyer of worlds', Fermi threw some pieces of paper in the air, and as the blast wave blew them away, he calculated the force of the bomb from the distance the papers flew. His result was not far from what the technical computations later came up with.

a shorthand for 'lost energy', which until this point is all that Pauli's idea really amounted to. If the neutrino really exists, it carries that energy along with it until it hits something. Fermi's theory had opened up a possibility for the neutrino to be revealed.

As the hero in H G Wells's *The Invisible Man* was detected by jostling in the crowd, so in Fermi's theory the phantasmal neutrino could hit an atomic nucleus, pick up electric charge, and turn into a visible electron. Were you to have been looking at the thing about to be struck, and been unaware of the neutrino, you might interpret the sudden movement of the target or the appearance of a high speed electron as a bewildering spontaneous creation of energy – the opposite of the puzzling lost energy of beta decay. Were the apparent energy shortfall in the accounts of beta decay to be matched precisely by the energy appearing in the target, the natural explanation would be that Pauli and Fermi were right: an unseen agent, created in beta decay, has transported energy across space until the carrier is destroyed and its energy passed on like a baton in some subatomic relay race.

So far so good. However, many a good idea dies as soon as the details are worked out. Fermi's theory said not just that a neutrino could pick up electric charge and reveal itself by bumping into matter, but predicted under what circumstances it would do so, and with what likelihood. This was where the difficulties began.

By 1934 there were enough data on beta decays from a range of elements that Fermi's theory could tell the overall chance of neutrino, electron, neutron and proton interchanging identities at a single point. It turned out to be trifling.[9] Hans Bethe and

---

[9] It is traditionally summarised in the 'Fermi constant', which is about one thousandth of a per cent when measured in units of the square of the proton mass.

Rudolf Peierls, two of the leading young theoreticians, realised that with this information and Fermi's theory, they could deduce the probability of interaction between neutrinos and matter, whereby a neutrino in flight might be exposed.[10] Hopes that Fermi's insight would lead to the neutrino's discovery were short-lived. Bethe and Peierls found that the chance of neutrinos revealing themselves in this way were puny. Their calculation implied that a neutrino produced in beta decay could travel through the whole Earth without interruption: 'like a bullet through a bank of fog.'

The interaction between a neutrino and matter became known as the weak force, once it was realised that a neutrino has a trifling chance of interacting with anything. Being electrically neutral, the neutrino does not respond to the electromagnetic forces that hold molecules together. Nor does it feel the strong forces that grip atomic nuclei. It only feels gravity and the weak force. The chance of a neutrino giving itself away by hitting a nucleus in some material was so small that the general opinion agreed with Bethe and Peierls' conclusion: 'There is no practically possible way of observing the neutrino.'

Had anyone other than Pauli proposed the existence of a particle that was effectively invisible, only revealing itself in the 'apparent' violation of energy conservation in an arcane nuclear process, Pauli might have dismissed it with his infamous critique, 'not even wrong'. Perhaps his wager of a case of champagne

---

[10] H Bethe and R Peierls, *Nature* vol 133, p. 532 (1934). Bethe and Fierz first calculated the chance from Fermi's theory and found it to be very small. Bethe and Peierls then used general principles to relate the chance of beta decay to the probability of a neutrino interacting with matter. Their result showed that this varies with energy but is always trifling at the energies relevant to the processes of interest at the time.

against anyone ever detecting the neutrino, thereby making a self-deprecating commentary, helped to deflect this criticism from himself. It looked increasingly as if Pauli had invented a piece of nothing that was gone without trace before you knew it, and even if you surrounded the site with prison walls made of lead a light year in thickness, the neutrinos would still have a good chance of escaping. The neutrino seemed to be a theorist's bad dream, a beautiful idea destined forever to be unknowable to experiment. In any event, the question of whether we would ever tease out a neutrino directly and prove its reality was forgotten, as physicists became embroiled in World War II. Nuclear fission, the outcome of Fermi's despair following his failed attempt to publish in *Nature*, filled their metaphorical radar screens. Pauli's wager would remain unchallenged for a quarter of a century.

WINNING THE LOTTERY

Bruno Pontecorvo

When Enrico Fermi gave up theorising about neutrinos in 1934 and started experimenting with neutrons, one of his collaborators was a young man named Bruno Pontecorvo. By bombarding atomic nuclei of various elements with neutrons, Fermi hoped to make new varieties of nuclei, or even elements. The products were invariably radioactive and it was by measuring this radioactivity that he hoped to weed out new products from the known and familiar. It was in the course of this that Pontecorvo, newly graduated from college and doing his first piece of serious research, noticed that when he moved the sample in its container, the amount of radioactivity seemed to vary, Fermi was intrigued, thought about it for a day and came up with the ideas that led to nuclear fission, the ability of neutrons in suitable circumstances to split the nuclei of heavy elements, and to liberate huge amounts of energy.[11] Within 10 years, these ideas were integral to the development of the atom bomb, and within 20 years were the fundamental seeds of nuclear power. Fermi won the Nobel Prize; Pontecorvo made a fortune from his share of the patents.

Laura Fermi's biography of her husband reveals not just the brilliance of the young Italian scientists, but also the social and political pressures they felt in Mussolini's fiefdom. She describes how Fermi was given permission to leave Italy to receive his Nobel award in Stockholm, never to return. He went to the USA where he played a central role in developing the atomic bomb. It was in 1954 that Laura Fermi wrote a biography of him. When I read it some 20 years later, I was struck by the story of Pontecorvo's disappearance.

Pontecorvo, having moved to Paris in 1936 to work with the Joliot-Curies, was unable to return to fascist Italy due to his Jewish background. He stayed in Paris, fleeing to the USA when the Nazis invaded. His strong socialist beliefs may be the reason why he was not invited to join the Manhattan project, and in 1943 he moved to Canada where he worked at the Chalk River Laboratory in Ontario. It was at Chalk River that he came up with the idea that would define the neutrino story for the rest of the century. We will come to that shortly, but it is what happened next that would prove to be so singular.

---

[11] It is not relevant for this story but in case you are wondering this is what was happening. It turns out that slow-moving neutrons have a vastly bigger chance of interacting than do fast neutrons. If, en route to the scene of their intended task, the neutrons from Fermi's source bumped into materials, and were slowed, this would increase the radioactivity that they subsequently caused. Materials with a lot of hydrogen atoms, such as water or paraffin wax, are the most effective at slowing neutrons (a consequence of neutrons and protons having the same mass). This trick of slowing neutrons and increasing their potency became a central feature in the operation of nuclear reactors.

In 1948, he took British citizenship and moved to the Harwell Laboratory in Britain. In post-war Britain, the atomic scientists were developing the H-bomb. The 'iron curtain' had descended over Europe and Klaus Fuchs, working at Harwell, was exposed as the 'atom spy'. Many intellectuals had become socialists in reaction to the rise of fascism in the 1930s, Fuchs among them, though by and large this did not extend to them sharing the results of their work with the Soviet regime. Fuchs however did.

In the USA, J Robert Oppenheimer, who had so singularly led the Allied teams building the first atomic bomb, was hounded from office on the grounds of his political opinions, and Senator Joe McCarthy led the notorious witch hunt against 'reds under the bed'. Fuchs's exposure increased the paranoia in the UK too. In the midst of this febrile climate, Bruno Pontecorvo disappeared.

On 21 October 1950, the newspapers carried the story, with speculation that he had slipped behind the Iron Curtain because he had the police on his tail. On 6 November, a statement was made in the British House of Commons by Mr Strauss, the Minister of Supply, to the effect that while there was 'no conclusive evidence of his whereabouts, [there are] no doubts that he is in Russia'. But no proof apparently. Three years later, Laura Fermi, one of Pontecorvo's oldest friends, would write 'over three years have now passed since the Pontecorvos' disappearance. No word has been heard from them. Nobody has seen them.'

And so apparently it remained, 20 years later as I read these words in the 1969 edition. Imagine my surprise therefore when, a few days after reading them, I saw a new paper about neutrinos in the scientific journal *Physics Letters*, written by one Bruno Pontecorvo, address: the Institute for Nuclear Research, JINR, Dubna, near Moscow. 'Did anyone realise?' I wondered. Of course they did, and had for a long time. He had won the Stalin Prize in 1953, and even given a press conference in 1955 explaining his reasons for leaving.

Stalin Prize, yes, but he never received the Nobel. As he died in 1993, he never will, but nine others already have, as a result of his ideas, and others may yet do so. During the course of our story, Pontecorvo will always be there behind the scenes, often a central actor and yet somehow never quite reaching the pantheon of the immortals. As we shall see, his self-imposed 'exile' in the USSR would later prevent some of his ideas getting the priority that they deserved. His first appearance had been as attendant in the aftermath of Fermi's theory of the neutrino as a player in beta decays. He made his first personal contribution to the neutrino story in 1946 by coming up with a way to capture a neutrino and prove its physical reality.

Everything that Bethe and Peierls had deduced from Fermi's theory for an individual neutrino produced in beta decay was true: the chances of detection were miniscule. But this was a statement of chance, and Pontecorvo realised that miniscule is not the same as nothing. It was while working at the Chalk River Laboratory in Canada in 1946 that he wrote his seminal report. The general belief before his paper was that detecting a neutrino is impossible. In Pontecorvo's opinion this seemed 'too drastic'. He believed that with 'modern experimental facilities' it might be possible. He then outlined his ideas on how to do it. For a single neutrino, think of yourself, and for 'miniscule' think of the chance of winning the National Lottery. I have never won it, and the chances are that you haven't either. Were there to be enough readers of this book, it is possible that one lucky winner might be among them, but for there to be a decent chance of a major prize winner, it would have to be top of the best sellers for many weeks – all of which is, regrettably, unlikely. The message though is clear: although neither you nor I are likely to win the top prize in the National Lottery, enough people buy tickets that someone beats the odds. The same is true for neutrinos: an individual neutrino produced in beta decay may travel the extent of the known universe without interruption, but if you were near an intense source producing billions of them each second, one or two might occasionally get caught in the atomic net.

Radium was the most powerful known source of beta decays, but even with large quantities of it, the numbers of neutrinos would be so small that the chance of capturing one would be hopeless. What was needed was some vastly more powerful source of neutrinos if there was to be any chance of detecting one.

Pontecorvo had been the one who had set Fermi on the road to developing nuclear power, and was working at a nuclear laboratory, so it is perhaps no surprise that he realised that the act of producing nuclear power in a uranium reactor should also be producing about ten million billion neutrinos each second. Pontecorvo realised that, with such vast numbers of neutrinos being spawned, with patience and the right detector it might be possible to catch a few. He then outlined his ideas on how this might be done.

When a neutrino hits a nucleus, Fermi's theory implied two things should happen. First, the neutrino picks up electric charge and turns into an electron. However, detecting this electron would be hopeless: there are electrons in everything and so it would be hard to distinguish one that had been knocked out of an atom from one created by a neutrino. It was the second implication of Fermi's theory that Pontecorvo homed in on: when a neutrino bumps into matter, the appearance of a negatively charged electron would be counter-balanced by an increase in the positive charge of the atomic nucleus that the neutrino had hit.

As the nuclear charge increased by one positive increment, it would be able to attract the negatively charged electron. The result of this is to make an atom of another element, the one placed one further rung up the periodic table of elements. Pontecorvo's insight was that if this atom was radioactive, it might be possible to detect its presence when it decayed.

Next, he outlined the requirements. The material used for detecting the neutrinos must not be too expensive as lots of it would be needed. The atomic nucleus produced by the collision must be radioactive, but not so much so that it would have decayed before this metaphorical needle in the haystack had been extracted. Also, extracting it would have to be easy if there was to be any chance of success.

These conditions gave the pointer to what would be best. He realised that if the target was liquid, and the element created by the neutrino collision was chemically inert, like helium, krypton or argon, there would be no danger of it reacting chemically, and so it could be extracted simply by boiling. Argon fitted the bill for the inert product, and next to it in the periodic table you find chlorine.

His idea was to use a huge vat of chlorine, in something cheap and easy to obtain such as cleaning fluid. If a neutrino hits the nucleus of a chlorine atom, the chlorine is transformed into an atom of argon. This argon atom is radioactive and decays, emitting radiation that can be detected with suitable instruments. If the vat of chlorine was large enough – hundreds of tonnes of cleaning fluid might be the solution – there was a chance of winning the lottery: a few neutrinos would hit, and radioactive argon be produced. The radiation emanating from the argon atoms would make them like radio beacons announcing that neutrinos had struck. That was Pontecorvo's insight. All it needed was someone with enough faith to take it on.

Enter Ray Davis

Ray Davis was born in Washington DC in 1914 and became interested in chemistry as a result of his father buying him chemicals for experiments in the basement. At home with chemicals, in all senses of the phrase, he took up the subject at university, getting his PhD at Yale in 1942. For the next three years he joined the war effort by testing chemical weapons, and at the armistice he joined the Atomic Energy Commission to work on radiochemistry – the chemistry of radioactive materials. He was by chance building up experiences that would soon help to forge his destiny.

In 1948, he joined the Brookhaven National Laboratory, on Long Island, New York, which was dedicated to finding peaceful uses for atomic energy. His first act on arrival was to talk to the chair of the chemistry department to find out what to do. Years later at the award ceremony for his Nobel Prize, this is how he recalled that meeting. 'To my surprise and delight I was advised to go to the library, do some reading and choose a project of my own, whatever appealed to me'. Thus began a long career 'doing just what I wanted and getting paid for it'.

In the library he came across a new review article about neutrinos.[iii] Several things were immediately obvious: very little was known, the field was wide open, and it was rich in problems. The seminal moment, Davis's epiphany, was the description of Pontecorvo's paper, which proposed a way of detecting the neutrino, which was well suited to someone with a background in

35

radiochemistry. The course of the rest of Davis's life was set at that moment.

Pontecorvo was proposing that chlorine converting to argon would be the signal. Davis knew that argon was an inert gas, easy to separate chemically from a large amount of chlorine solution. The particular atoms of argon produced in this way would be radioactive, decaying with a half-life of 35 days as they revert back to chlorine. Davis knew how to detect radiation by its ability to ionise gas molecules, giving rise to electrical signals.[12]

For Davis, this seemed almost too easy. And so it would prove, but for no fault of his or of Pontecorvo.

Brookhaven had a modest test reactor of its own on the site, which was used for research. He set up a tank containing 4000 litres of carbon tetrachloride next to this and waited for enough argon to accumulate. Then he went through the procedures and found nothing other than the result of impacts from cosmic rays. The signals, such as they were, were no bigger when the reactor was operating than when it was not. So in 1955, he built a larger detector and took it to the newly opened Savannah River nuclear reactor in South Carolina.

Here too the result was the same: nothing. What no one then knew was that the nuclear reactions were primarily producing not neutrinos but *anti*neutrinos.[13] Just as the electron has an antimatter doppelganger, the positron, and the proton is mirrored by an antiproton, so do all varieties of matter have their antimatter analogue. Neutrinos and antineutrinos are

---

[12] This is the principle of a gas-filled proportional counter, the modern version of a Geiger counter.

[13] The decay $n \rightarrow pe^-$ involves one particle of matter at the start and two at the end, so to balance the accounts an antineutrino is required in the decay products.

like Tweedledum and Tweedledee; chlorine would be fine for detecting neutrinos, but to detect antineutrinos you would need tanks full of antichlorine. Later, it would be realised that this failure was, in a way, a triumph: Davis had implicitly proved that neutrinos and antineutrinos are different. But at that time no one had proved that the neutrino exists, and Davis was deflated by seeing – nothing.

Pontecorvo's idea was correct. A tank of chlorine *would* be an ideal way to capture neutrinos. And had reactors been producing large numbers of them, Davis would surely have discovered the neutrino by 1955. One could imagine that had he had a tank of antichlorine he might indeed have discovered the antineutrino, but antimatter in bulk is science-fiction.[14] Fortunately, there are other ways of capturing an antineutrino, but Davis would have to wait for a source of neutrinos for his chlorine detector to come into its own.

Project Poltergeist

Towards the end of World War II, Fred Reines joined the Manhattan project at Los Alamos. In 1944, he first became group leader in the theoretical physics division at the laboratory, and then the leader of Operation Greenhouse, which consisted of a number of Atomic Energy Commission experiments on the Eniwetok Atoll. He worked on the results of bomb tests there and at the Bikini atoll, and at the Nevada testing grounds. His main efforts were in understanding the effects of nuclear blasts.

---

[14] Antimatter, in fact and fiction, is the theme of my book *Antimatter*, Oxford 2009.

The idea of seeking evidence for the neutrino had occurred to him after reading Pontecorvo's theoretical paper in 1947, but he had no opportunity and didn't pursue it. It was in 1951, when he was on sabbatical leave and thinking about physics that he might do in the coming years, that the idea returned. He later recalled that he 'moved to a stark empty office, staring at a blank pad for several months searching for a meaningful question worthy of a life's work'.[iv] His sole inspiration came out of his experience with atomic explosions. Atomic bombs give off lots of neutrons and when these decay they produce neutrinos (or, as we now know, antineutrinos). This offered the chance that out of these hordes, some 'neutrinos' might reveal themselves, if only rarely.

He did some rough calculations and decided that all he required was a small detector, about a cubic metre in size. What he really needed was an expert to consult.

During the summer of 1951 Enrico Fermi visited Los Alamos. Reines realised that having the great man working in an office a few doors away was too good an opportunity to miss, so he plucked up courage and went to ask him about neutrino detection. Fermi agreed that using a bomb as the source was best. Reines felt 'so far, so good' and then admitted his problem: he had no idea how to make a suitable detector. Fermi thought about it for a while and then confessed that neither did he. Reines was deflated and forgot about it until he had a chance conversation with Clyde Cowan.

He and Cowan were flying to Princeton when the plane was grounded in Kansas City with engine trouble. Wandering around the airport they started to discuss what might be the most difficult experiment in all of physics. Cowan suggested a problem in atomic physics,[15] but they decided that others had already started to work on that. Then Reines suggested that they should focus on the neutrino. Cowan immediately replied, 'Great Idea!'

Although a nuclear explosion may be a great source of neutrinos, it has problems. The idea of having a sensitive detector within 100 metres of the most violent man-made explosion ever was somewhat bizarre. However, they had both worked with bombs and were confident that they could protect the detector by placing it underground. The director of Los Alamos gave permission for them to go ahead.

Exploding an atomic bomb is a one-off event, and so it would be critical to be sure that they had everything under control. In particular, Hans Bethe asked whether they could be certain that they could distinguish a genuine neutrino from other radiation emitted by the bomb, such as gamma rays and neutrons. It was in the process of coming up with answers to this that in September 1952 they realised that there was a better way to do the experiment. Controlled nuclear power, in the form of a nuclear reactor, would work equally well as the source.

A nuclear reactor typically would emit ten trillion neutrinos per square centimetre per second, which ought to be enough. Reines later said that he wondered why it took so long for them to come to 'this now obvious conclusion, and how it escaped others'. They had no worries about being scooped because 'neutrino detection was not a popular activity in 1952'. They wrote to Enrico Fermi on 4 October telling him that 'only last week it

---

[15] This was to measure the properties of positronium, where an electron is bound to a positron. Being matter and antimatter the electron and positron annihilate such that positronium lives for less than a millionth of a second – see *Antimatter*, Frank Close.

occurred to us that we could do [the experiment] at a nuclear reactor', and asked him for comments. He was obviously smitten because he immediately replied, agreeing that it would be much simpler, and adding the cogent remark that an experiment at a reactor would 'have a great advantage that the measurement can be repeated any number of times'.

The (anti)neutrinos from a reactor can induce a process known as 'inverse beta decay', where an antineutrino hits a proton, converting it to a neutron, the proton's electric charge being carried away by a positron, the antimatter version of an electron. Cowan and Reines were unaware of the subtle distinction between neutrino and antineutrino – that was still in the future – but they did know that if 'neutrinos' exist, the conservation of electric charge would make the products be a neutron and a positron. That would be enough for their scheme to work; relative to Davis, they were fortunate or inspired – take your pick.

They built a small prototype detector in 1953 at a nuclear reactor at the Hanford Engineering works in Washington State. They named it Project Poltergeist because of their quarry's ghostly nature, and had the first hints of a signal that year.[v] However, any excitement was only temporary because they continued to measure signals even when the reactor was switched off! Davis and they were experiencing similar frustrations, though at this stage none of them knew of their rival's efforts.

Cowan and Reines realised that at Hanford their experiment could not be shielded from cosmic rays, and collisions between these and atoms in their detector were giving signals that mimicked those they were looking for. Although they felt that identification of a free neutrino had been made, they needed a better experiment to carry it to a more definite conclusion. Their story at this point continues to have uncanny parallels with that of

Davis – as in his case, they too built a larger version of the detector and, in 1955, took it to Savannah River. Here they could locate it 12 metres underground, well shielded from cosmic rays while being less than 11 metres from the centre of the reactor.

The idea of Poltergeist was to detect two separate bursts of gamma rays, light far beyond the visible spectrum, which should occur 5 microseconds apart from one another if an (anti)neutrino had been captured. The immediate result of such an event would be the appearance of a positron and a neutron. The positron would annihilate almost instantaneously with the ubiquitous electrons, present in everything. This would produce two gamma rays. The second burst would come when the neutron was captured by a nucleus of cadmium atoms in tanks of cadmium chloride. To be captured, the neutron would have to have slowed by successive collisions, and this would take about 5 microseconds. Hence the two separate bursts of gamma rays.

And that is exactly what happened. The cosmic ray background was minimal and overwhelmed by the radiation from the reactor. In the summer of 1956, 'Poltergeist' recorded gamma rays bursts separated by 5.5 microseconds. On 14 June, Cowan and Reines sent Pauli a telegram announcing that they had finally found the neutrino that he had invented a quarter of a century earlier. The news was flashed around the world. One of my first memories of physics was hearing on the radio that year that 'the neutrino has been discovered'.

Years later Reines reminded Bethe about his pronouncement with Peierls in 1934 that 'there is no practically possible way of observing the neutrino'. With a smile, Bethe replied 'Well, you shouldn't believe everything you read in the papers.[16]

---

[16] As recalled in F Reines's Nobel Address.

After the discovery, Reines devoted his entire career to understanding the properties and interactions of neutrinos. In 1995, he won the Nobel Prize recognising the several discoveries that he had made during those 40 years. But Cowan had died in 1974, and many feel that recognition for this discovery should have been made years earlier. However, they did win from Pauli the case of champagne that he had wagered so long ago. Ray Davis and Pontecorvo would have to wait.

4

IS THE SUN STILL SHINING?

Even on the gloomiest overcast day no one doubts that the Sun is still there. 'Is the Sun still shining?' asks whether it is still showing between gaps in the clouds, not whether it has terminally quit. But in the 1970s, some scientists briefly contemplated the possibility that the Sun's fuel was exhausted, that the visible glowing orb was just its dying embers, and that the ultimate energy crisis had begun.

Sunlight takes only eight minutes to reach Earth, so its brilliant surface shows that at least its outer limits were still shining very recently. The temperature of this visible bright region of the Sun is about 6000 degrees, hotter than a blast furnace, but not unimaginably so. The dazzling intensity hides its inside from view.

The origins of this light lie deep within. Energy from the core rebounds for thousands of centuries before surfacing; what you see today is the end product of reactions that occurred more than

100,000 years ago. If the heart of the Sun has already burned out, it could be some time before we see it dimming. And that is what started troubling scientists 30–40 years ago.

Here are some of the ideas that were put forward. In 1973, Andrew Prentice suggested that the Sun had burned out leaving a core of helium; Fred Hoyle in 1975 suggested that its core contained a lot of heavy elements that had survived the Big Bang and attracted a halo of hydrogen five billion years ago as the solar system was being formed; and also in that same year another group of theorists suggested that there was a black hole at its centre.[vi] In 1980 I wrote an article in *Nature* about this with the question 'Is the Sun still shining?' as its title.[vii] This captured the attention of the BBC television's flagship science programme of the time, *Tomorrow's World*, and the story spread around the world.

What was the reason for all this fuss? Why does the Sun shine at all? And what do neutrinos have to do with any of this?

Great Ball of Fire

How the Sun has produced so much energy day in day out for the entire time that the Earth has existed is one of the oldest questions in science. Charles Darwin even began to doubt his *Origin of the Species* because nothing in physics or chemistry 150 years ago could explain how the Sun and the Earth could have lasted long enough for the vast time spans that geology and his theory of evolution would have required. It is only in the 21st century that the answer has been finally proved.

The ancient Egyptians thought that the Sun was a ball of fire. This was a natural extrapolation from their limited experience, though they don't seem to have had any opinion as to what its fuel consisted of. The Greek philosopher Anaxagoras in the 5th century BC was the first to come up with a theory.

'Shooting stars' are lumps of rock that have hit the Earth's atmosphere at speeds of kilometres each second. They become red hot through friction as they fall towards Earth. Some are large enough that their remnants – meteorites – reach the ground. Two and a half thousand years ago, Anaxagoras found one that had just landed. It was a lump of metal, and still so hot that he decided that it must have come from the Sun. This gave him a sudden flash of inspiration: the sun must be made of red hot iron. This was the widely held view for almost two millennia.

It is less than 200 years ago that the first problems began to emerge with this simple picture of the Sun.

The industrial revolution was under way, and with it came a growing understanding of thermodynamics and the significance of the conservation of energy. Fuel had to be supplied continuously to keep the blast furnaces working. If left to itself, molten iron in the steelworks would rapidly cool, and scientists realised that even the Sun could not stay hot for ever. It was during this same period that discoveries in geology, and the arrival of Darwin's theory of evolution, independently pointed towards a common message: the Earth had to be more like hundreds of millions rather than thousands of years old. Yet the known laws of physics could not explain how the Sun could have burned so long.

John Waterstone, a schoolteacher, around 1850 showed that chemical energy could have fuelled the Sun only since the stone age, some ten thousand years ago. As the Sun had existed much longer than that, some other power source must be at work. The only possible candidate then known was the force of gravity. As material such as rocks fall under the gravitational force of the Sun, they gather speed. Upon hitting the Sun and stopping, this energy is turned into heat, similar in principle to the heating of a car's brakes and tyres when the car is suddenly brought to a halt. Waterstone suggested that meteors falling in from space hit the Sun and, in doing so, produced the heat that powers the solar furnace.

Having come up with this clever idea, he did the sums and realised that there are too few meteors to do this. So he refined the idea, and proposed that the Sun itself was falling inwards under its own weight, producing heat. He spoke about this at the annual meeting of the British Association for the Advancement of Science in 1853. William Thomson was in the audience and was clearly impressed. Thomson developed the idea into its sharpest form, and found that even here the sums did not add up. He also considered the possibility that the Sun would consume the planets one by one, but this didn't work either. Were it to capture Mercury and Venus, these would only power the Sun for a century, and consuming all of the planets would only give it 3000 years of life.

In 1860 Thomson took up the idea of the shrinking Sun once more, contemplating how it might collapse while producing heat for the longest period.[17] In an article in *Macmillan's Journal* in 1862 Thomson concluded that the Sun's age was most probably 'not more than 100 million years'.

A finite timespan for the Sun implied both future apocalypse and a limit to history, with implications for Darwin and for geologists who were looking at timescales longer than this. In the later edition of his book, *On the Origin of Species*, Darwin removed all mention of timescales. As early as 1869 he had been so shaken

---

[17] Waterstone had already thought about collapse too – the full story is in John Gribbin's book *Blinded by the Light*.

by Thomson's analysis that he had written to Alfred Russel Wallace, co-discoverer of natural selection, that the implications of Thomson's work for the age of the world 'have been for some time one of my sorest troubles'.

Years later, in 1897, Thomson (by then Lord Kelvin) refined his calculations and announced that the most likely age of Sun and Earth was about 25 million years, which made the paradox even starker. Thus evolution, geology and physics were in conflict unless, as he prophetically added, 'there are sources now unknown to us in the great storehouse of creation.'

When making these calculations, he had implicitly assumed that matter in the Sun and the Earth were similar. Thus it is ironic that, in 1897, another Thomson, J J Thomson (no relation) had discovered that atoms have an inner structure – they contain electrons. This meant that atoms might differ in the searing heat of the Sun from those on Earth. In turn it was also possible that they followed new laws. In addition, the discovery of radioactivity by Becquerel, the previous year, indicated that nature has the means of spontaneously producing radiant energy by as yet unidentified means. Yet neither Lord Kelvin nor anyone at that moment seemed to have put these possibilities together to reconcile the paradoxes.

A New Source of Energy

Although Becquerel had discovered radioactivity, no one initially regarded it as especially important. During the final decades of the 19th century a whole range of weird radiations had turned up, such as fluorescence and X-rays, so 'Becquerel rays' appeared to be just another to add to the list. It was only when

Ernest Rutherford got to work that the full implications of the power latent within the atom, of which radioactivity was but a part, would become known. Rutherford's name will forever be associated with the unravelling of the structure of the atom. About the only thing that he got wrong in the next 30 years was his judgment that anyone thinking there was useful energy within the atomic nucleus was talking moonshine. In reality, nuclear energy is the source of sunshine.

In their different ways, all of Rutherford's discoveries would lead towards the neutrino. The most immediate for our tale is his first result: disentangling the nature of radioactivity.

As we have seen, Rutherford identified different forms of radioactivity. The alpha form consisted of particles that turned out to be the nuclei of helium atoms. This explained a puzzling observation: that traces of helium gas had been found in minerals that contained uranium.

Radioactivity was as powerful down dark mines as in daylight, which showed that the energy must be coming from within the atoms themselves. Rutherford had also found that elements transmute one into another in radioactivity. Putting all this together gave him the essential insight: the continuous emission of energy in radioactivity comes from the changes in the internal structure of the atoms. Rutherford first showed that the more alpha particles were radiated, the more energy was emitted. Natural radioactivity was found in the air and the rocks beneath our feet. It turned up everywhere. The idea that radioactivity could warm the Earth began to take hold after radium was discovered. Radium is naturally warm and continues to emit heat without cooling down to the temperature of the surroundings. Here was the first hint that Lord Kelvin's calculations of the Earth's age, based on thermodynamics, might not be the full story.

Rutherford realised that radium's continuous heat output was evidence for a new source of energy, and that Kelvin's paradox could be avoided. He announced this at a talk at the Royal Institution in May 1904. The room was gloomy and he spotted Lord Kelvin in the audience. He recalled that he would be in for trouble when he came to the final act of his speech, which dealt with the age of the Earth, on which their views differed radically. To Rutherford's relief Lord Kelvin fell asleep until the critical moment when, Rutherford later recalled, 'I saw the old bird sit up, open an eye and cast a baleful glance at me'. Inspiration arrived on cue however, with Rutherford announcing that Kelvin had limited the age of the Earth '*provided* no new source of energy was *discovered*. That prophetic utterance refers to what we are now considering tonight: radium! The old boy beamed upon me.'[viii]

Kelvin had assumed that the planet is simply a cooling body, but Rutherford's point was that radioactivity supplies heat within the Earth continuously. As the planet has this internal power supply, its age can be far greater than Kelvin had computed. Today the rate that elements, such as uranium, decay by radioactivity has been carefully measured. This shows that half of a sample of uranium decays on a timescale of about 4.5 billion years.[18] By measuring the rates for other elements, and then comparing the relative amounts found in minerals, it is possible to determine how long has elapsed since the original sample was formed. This places the Earth at some 4.5 billion years old, in line with the geological and evolutionary evidence. Meteorites have also been found with ages around 5 billion years. This all

---

[18] The two main isotopes of uranium, U-238 and U-235 have half-lives respectively of 4.5 billion and 700 million years.

fits with the notion that the solar system began about five billion years ago, and that the Sun has burned for that time.

While the warmth from radioactivity explains the age of the Earth, and this novel energy source suggests a way for powering the Sun, it is still a long way from here to solving exactly how the Sun does it.

The Nuclear Sun

As radioactive elements can emit energy continuously, the natural first guess was to suppose that the Sun is made of radioactive elements. In 1903, William Wilson, an English astronomer, calculated that just a few grammes of radium per cubic metre of the Sun would be enough to explain its power. Here was the first proof that solar power and radioactivity could be linked with reasonable amounts of fuel being needed. However, this idea was quickly ruled out as no sign of radium showed up in the solar spectrum. Whatever was responsible for solar power, it was not radium.

The element most famously associated with sunlight is helium. Helium was named after Helios, the Sun god, as it had been discovered in the spectrum of the solar atmosphere and clearly is present in the Sun. However, helium is not radioactive, though it is produced in radioactivity, alpha particles being the nuclear seeds of helium atoms. This stimulated the idea that helium might be the residue, or ash, of the primary power production in the Sun. However, there was a problem: there was no sign in solar spectra of any of the heavy radioactive elements known to produce alpha particles. Whatever helium's origins are in the Sun, they are not from radioactivity as on Earth.

The major breakthrough came two years later when Einstein's theory of relativity appeared with its famous '$E=mc^2$', and the implication that all forms of matter are latent with energy. Einstein pointed out that if a body emits an amount E of energy, its mass m reduces by an amount E/c^2 where c is the velocity of light. Rutherford, in 1913, by then having discovered the atomic nucleus at the heart of every atom, commented that at the enormous temperatures in the Sun, elements that on Earth appeared stable might behave very differently, changing from one variety to another. In so doing, some of their mass could convert into radiant energy, whereby the mass of the Sun would in consequence be falling over the aeons. This isn't as startling as it seems. Nuclear transmutation gives a lot of energy for just a trifling amount of mass, because the speed of light, the c in the $E=mc^2$, is so large. Were just one per cent of the Sun's mass turned into energy, it could burn for a trillion years. The idea that the Sun is a nuclear furnace was born, but several years were to pass before the way it actually works was explained.

The modern theory of sunlight began in 1920 with an insight by Sir Arthur Eddington, Professor of Astronomy at Cambridge. As we saw, Rutherford had already suggested that at enormous temperatures transformations might take place among elements that we on Earth regarded as stable. Although he did not say so explicitly, this laid open the possibility that helium in the Sun might be produced as the result of a nuclear transformation that was unknown on Earth. This was what Eddington was now proposing: the Sun generates heat and light by burning hydrogen and turning it into helium.

He had the idea as a result of a discovery made by his colleague Francis Aston at the University's Cavendish Laboratory. Aston had found that an atom of helium has one part in 120 less mass than four atoms of hydrogen. That is what gave Eddington the clue. Could the helium that had been discovered in the solar atmosphere be some of the ash left over when a Sun made of the simplest element of all, hydrogen, converts the hydrogen into helium? Every time four protons – the nuclei of hydrogen atoms – fuse together in the Sun's core, they can make a single atom of helium and the '1 in 120' mass loss is converted into radiant energy. This is Einstein's $E = mc^2$ at work.

The obvious questions are why does this not happen on Earth, and what is special about the Sun?

There is a fundamental property of electricity that particles carrying the same sign of electric charge will repel one another. All protons carry positive electric charge and so feel this resistance to pairing. In order to force two protons together and start building up the seeds of helium, this repulsive force must be overcome. At earthly temperatures, protons are effectively stationary and have no chance of bumping into one another; as a result, hydrogen does not naturally fuse to make helium in the cold.

Stars, such as the Sun, are very different from the Earth. Their stability is a balancing act between the tendency for the star to collapse under its own weight and the ever-increasing thermal violence near its core. The temperature near the centre of the Sun was calculated to be about 14 million degrees if its stability, size, mass and brightness were to be explained. At such temperatures the protons are moving fast, and can get close enough to one another so that they can join, or fuse, before the resistive electrical force has slowed and rejected them. The result is that in the centre of the Sun, protons can stick to one another and turn into helium.

Eddington had suggested how the Sun produces its energy but not the details of how it takes place. The full theory only arrived in 1939 when Hans Bethe published his paper 'Energy Production in Stars'.[19]

According to legend, Bethe solved the problem during a train journey from Washington DC to Cornell in upstate New York. Astronomers and physicists had gathered in Washington for a discussion about how stars produce energy. Bethe was a young nuclear physicist, and already an expert on nuclear transmutation, but until he went to that conference he had been unaware of the interest in stellar energy and the general opinion that nuclear processes had to be involved. The challenge was to find examples of nuclear transmutations that would produce energy high enough to maintain the Sun's energy output, but not so fast that the Sun would self-destruct. The goal therefore was to find the 'Goldilocks' reactions – the ones that were not too hot, not too cold, but just right to keep the furnace steadily burning at 14 million degrees.

With the confidence of youth, Bethe decided that this should be an easy problem to solve, given his nuclear expertise. So he set himself the challenge of doing it on the train journey home, preferably before the stewards called passengers in to dinner. He succeeded in discovering the CNO cycle,[20] which was quite different from what Eddington had suggested. 'CNO' stands for carbon, nitrogen and oxygen, and the cycle occurs when protons in a star containing some carbon can convert it first into nitrogen, then to

---

[19] Between Eddington's idea in 1920 and Bethe's work in 1939, the Welsh scientist, Robert Atkinson, and the Dutch, Fritz Houtermans in 1929 had used the measured masses of the light elements, together with $E = mc^2$, and predicted that large amounts of energy could be released by fusing the nuclei of light elements together.

[20] His paper published in 1939 – *Physical Review*, vol. 55, p. 434 – was received by the editor in September 1938. That same year, the German Carl von Weisacker had independently also realised the possibility of the CNO cycle.

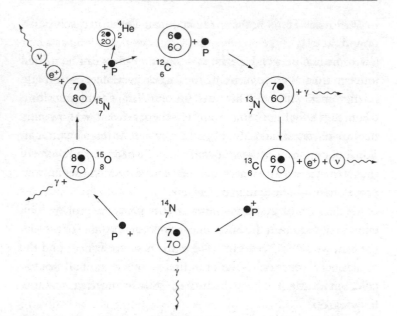

Figure 3 The CNO cycle. The CNO (Carbon Nitrogen Oxygen) cycle is the dominant source of energy production in stars that are heavier than the sun. The result is the fusion of four hydrogen nuclei (protons) to form a single nucleus of helium, denoted He. The nuclei of various elements are denoted by symbols H He C N O; the superscripts denoting the total numbers of constituents (protons and neutrons) and the lower denotes the number of protons. The γ (gamma) denotes a photon and ν a neutrino. This process produces only relatively low energy neutrinos. The wobbly lines illustrate the emission of energy as photons and neutrinos from the star. Protons are denoted by solid dark circles; neutrons by open white circles.

oxygen and then back to carbon again by emission of helium (Figure 3). This was a beautiful theory, producing energy and helium: helium – the element that had been discovered in helios, the Sun. It looked perfect.

With pen and paper in a railroad car, Bethe had solved an important part of the puzzle of how stars work. So long as a star has some carbon as a catalyst, any spare protons can be turned into helium and power, leaving the carbon available to stimulate further such fusions. This was fine but begged the question: where did the carbon come from? His theory explained how stars that are hotter than the Sun, and about half as large again, can shine, but he soon realised that it did not work for the relatively smaller, cooler Sun, where carbon, as we now know, is rare. Some other process had to be involved.

At Cornell University, Bethe decided to study the problem systematically, working through the entire periodic table of the elements if needs be. Thankfully he didn't have to, as he found the solution at the very start with the simplest element of all: hydrogen. In so doing he had rediscovered Eddington's idea, but now he worked out the consequences, turning it into a quantitative description, amenable to scientific test.

At temperatures of millions of degrees, as in the centre of stars like the Sun, atoms of hydrogen are ripped apart into their components: electrons and protons. When protons bump into one another there is a chance of nuclear reactions – fusion – taking place. What has become known as the '*pp* chain' (*pp* for proton–proton) begins with a collision between two protons where they fuse together forming a deuteron (a loose system of a proton bound to a neutron), a positron and a neutrino.[21] The deuteron finds itself in a crowd of protons, and almost immediately grabs one; the resulting trio is a nucleus of helium-3, consisting of two

---

[21] One of the protons having converted to a neutron, positron and neutrino, as in the form of beta decay discovered by Joliot, page 20, and described by Fermi's theory.

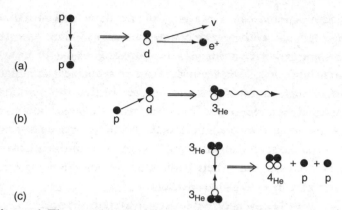

Figure 4 The *pp* proton-proton chain reaction in the sun. Two protons denoted *p* fuse to make a deuteron (consisting of a neutron, the white circle, and a proton, the dark circle) together with a positron (e^+) and neutrino. In (b) another proton hits the deuteron, converting it to helium-3 and a photon. In (c) we see the consequence of two of these processes: two nuclei of helium-3 combine to make one of helium-4 and two protons.

protons and a neutron. Finally, when two nuclei of helium-3 collide, they form the stable form of helium, helium-4, and throw off two protons. The net result is that four protons at the start have ended up as a single seed of helium-4, emitting energy in the form of positrons, photons and neutrinos (Figure 4). Whereas the CNO cycle needs temperatures above 20 million degrees to be effective, the *pp* chain works at 15 million degrees, as in the heart of our Sun.

The positrons annihilate with electrons, producing gamma rays – particles of light far beyond the visible spectrum. Electric charge is like a barrier to photons, which grabs them, soaks up some of their energy and then throws them off again. These photons take thousands of centuries to bounce their way upwards to the surface, where they emerge as light that is visible to our eyes. The neutrinos stream out unimpeded, reaching Earth in a little over eight minutes. At least, that is what would happen if Bethe's theory of starlight were correct. The numbers balanced, and the physics made sense, but only experiment would be able to tell if it were actually true.

Experiments on Earth have shown us how protons behave. When two of them collide at speeds similar to those that of particles in a gas at 14 million degrees, it turns out that they can encroach near enough to fuse. Even so, it is very unlikely: only about once in ten billion trillion encounters are two protons likely to meet and fuse to make a deuteron, initiating the *pp* chain. Put another way, if you were a proton in the Sun, after 5 billion years it is roughly a 50:50 chance as to whether you would have found a partner and fused. This is rare to be sure, but there are lots of protons that make up the Sun. About 5 million tons of mass, in the form of protons, is being converted into helium each second, the energy released giving warmth to the Earth and light visible across interstellar space. This slowness is good news, for it has enabled the Sun to burn long enough for the evolutionary processes leading to intelligent life to have occurred here. It solves the conundrum of how the Sun has shone for so long, and it fits with the five billion year age for the Earth and the solar system.

So Bethe's genius had finally found nuclear processes that could produce heat at the right rate to explain the Sun as we see it, and also to produce helium as the ash. We know the size of the Sun, its mass, and we see its light. The sums balanced. His explanation fitted the known facts, which is the critical first requirement, but it still begs the question: is this what *actually* takes place in the Sun? Critically, the hypothesis had further implications that no one had ever tested before: Bethe's theory of nuclear fusion in the *pp* chain implies that neutrinos are produced.

However, when he proposed the idea in 1939, the neutrino itself was still only a figment of theorists' imaginations. Reaction to Fermi's theory had shown (page 24) that the neutrino was regarded as 'remote from reality' and even 'not of interest'. Bethe's own calculations with Peierls in 1934 had shown that the neutrino, if it existed at all, had such a small chance of hitting anything that it would most probably be unobservable. This remained the received wisdom until Pontecorvo's paper in 1946.

This may be why Bethe's paper made no mention of the possibility of testing the theory by detecting neutrinos from the Sun. It was only after Pontecorvo came up with his ideas and the neutrino[22] was discovered in 1956, that the possibility of testing Bethe's idea by looking for solar neutrinos began to take hold.

---

[22] Actually antineutrino, see page 36.

HOW MANY SOLAR NEUTRINOS?

Early Ideas

We only ever get a superficial view of the Sun as we are blinded by the light to what goes on hundreds of thousands of kilometres beneath, in the core of its nuclear furnace. But if Bethe was right, neutrinos are pouring out unhindered. If we could capture neutrinos from the Sun, we would in effect be looking into the heart of a star. That is what inspired Ray Davis.

To succeed in finding something, it helps first to have an idea of what you are looking for. How much energy does an individual neutrino have? Knowing this, would tell us what sort of detector would work best. Is the Sun bright in neutrinos, or dim? This would determine how powerful, how large, the apparatus would have to be. Might the solar neutrinos be so dim that it would be utterly impossible to detect them? Having answers to such questions would be crucial. The way that I am describing this may give the impression that a clear managerial plan unfolded from Bethe's theory of the Sun, for designing a detector, and finally seeing solar neutrinos. The reality was far from that. Indeed, hardly anyone gave it much thought.

Bethe's idea, that nuclear fusion is the basis for energy production in stars, had first emerged in 1938/9, five years after Pauli and Fermi had presented their theory of beta decay, which included the neutrino. Even though Bethe's theory implied that the Sun should be producing not just heat but vast numbers of neutrinos, none of the early papers on nuclear fusion in stars mentioned the possibility of testing the idea by detecting them. Bruno Pontecorvo's paper in 1946, which had inspired Ray Davis's unsuccessful attempt to catch neutrinos coming from a reactor, only mentioned the Sun in two sentences. Recall that his main point had been that chlorine could be a good detector of neutrinos – so long as there were large numbers of neutrinos and you had enough chlorine.

Although Pontecorvo had only made passing mention of the Sun, and the review article that had first excited Davis's interest had also only included the idea briefly, nonetheless it had attracted his attention. Even as he was making his first attempt at catching neutrinos from the reactor at Brookhaven, he also realised that his apparatus might capture solar neutrinos – if the CNO cycle was important.

Chlorine is only an effective detector if the neutrino has enough energy to induce the reaction[23] that provides the crucial evidence

---

[23] The energies are usually written in electron-volts, eV. This is the energy that an electron would gain when accelerated by a one volt battery. The amount of energy released in a single atom in a chemical reaction is usually a bit less than this. In nuclear processes thousands (keV) or millions (MeV) of electron-volts are exchanged. A neutrino needs at least 860 keV to activate an atom of chlorine and convert it to the radioactive form of argon that Davis could detect. Only one in ten thousand neutrinos from the Sun has this much energy.

by changing atoms of chlorine into argon. According to Bethe's theory the pp chain is most important in the Sun, whereas the CNO cycle is dominant only in larger stars. Unfortunately, the pp fusion process produces neutrinos whose energies are less than half that required to affect the chlorine; in effect, a chlorine tank is blind to them. However, when Ray Davis first became interested in the challenge, astrophysicists were still debating whether the CNO cycle might play some role in the Sun. In this cycle, the production of nitrogen-13 and oxygen-15 produced neutrinos with sufficient energies to trigger the chlorine detector.

He also worried about backgrounds – random effects that could disturb the chlorine in a similar way to neutrinos, and give false signals: not every burglar alarm reveals a malicious intruder. For this reason, in 1955, Davis buried his 4000 litre detector six metres below the soil in order to reduce the background from cosmic rays. As we saw on page 36, his attempt to capture neutrinos from the reactor was hopeless, because a reactor produces intense amounts of *antineutrinos*. At least the Sun was predicted to produce the genuine article, so there was hope that this time he would be successful.

Unfortunately, he had no better luck here either. After several weeks it was becoming all too clear that he hadn't found any evidence for neutrinos from the Sun. He wrote up his report announcing that if the Sun's power was produced by the CNO cycle, then his failure to see any neutrinos meant that the production rate must be less than a rather small number. In effect, either the idea of solar neutrinos was wrong, or the CNO cycle wasn't important.

His experimental renunciation of the CNO cycle did not raise much interest because by then astrophysicists had become convinced that the Sun was powered primarily by the pp chain and *not* by the CNO cycle. One reviewer of his paper criticised it on several grounds. First, as there was at that time (1955) no evidence that the neutrino even existed, Davis's failure to find any didn't imply much about the Sun. Second, even if they were being produced this way, Davis's experiment did not have enough sensitivity to detect any, as the CNO cycle is trifling in the Sun anyway. The enterprise was compared to someone standing on top of a mountain, reaching for the moon, and, when failing to touch it, concluding that the moon is more than three metres from the top of the mountain.[ix] In a nutshell: underwhelming.

A Glimmer of Hope

In 1958, there was a dramatic discovery about the nuclear processes that Bethe believed powered the Sun.

In the *pp* chain that produces helium, the final stage produces helium-4, after two nuclei of helium-3 collide. If there were nothing but protons in the Sun to begin with, this would be the whole story. However, the Sun has been doing this for five billion years, so there is already a lot of helium-4 in there as well. At this very moment, the *pp* cycle is making helium-3 and there is a chance that this will not meet another newly made helium-3 but instead will bump into some of the historic helium-4. When this happens, helium-3 and helium-4 combine to make a nucleus built of seven constituents, four protons and three neutrons: the nucleus of beryllium-7.

Bethe had realised that this production of beryllium could happen, but only rarely. Then at the annual meeting of the American Physical Society in New York came the news that scientists in an experiment at the Naval Research Laboratory had managed to fuse these two nuclei, helium-3 and helium-4, together and found it to be a thousand times easier than anyone had suspected. This meant that their fusion would happen in the Sun a thousand times more often than had been previously thought.

Willy Fowler, one of the leading astrophysicists, had travelled from his home base, Caltech in Los Angeles, to the meeting. The announcement triggered a memory about something that Bethe had once remarked to him. It was that if beryllium-7 was produced, it could bump into one of the solar protons, these fusing together to make a clump of eight: boron-8. The exciting thing was that in the process of doing so, a neutrino would be emitted with energy well above the threshold needed for triggering a chlorine detector. The Naval people's results were implying that beryllium-7 would be produced much more often than Bethe had thought and, Fowler realised, neutrinos also. Suddenly a tantalising possibility of detecting solar neutrinos had presented itself. Fowler immediately wrote to Davis that the Sun might indeed be producing neutrinos in large enough numbers for a chlorine detector to capture.

When Davis heard the news, his experiment at the Savannah Reactor was already underway. There was no doubt that argon atoms – the signal for neutrinos – were being produced, but the amounts stubbornly fitted those expected to come from interactions of cosmic rays. The experiment was not sensitive enough to separate any solar neutrinos from this unwanted background. It was as if Davis was trying to tell if it had started raining, while situated underneath a waterfall: if there were any solar neutrino 'raindrops' they were lost in the cosmic cascade. If he was going to do better than his 1955 failure, the effect of the cosmic rays would have to be reduced. As one scientist pithily summarised things,[24] astronomers go to the top of mountains to get away from the glow

---

[24] Art McDonald, head of the SNO Collaboration, chapter 9, used this analogy in a press conference in 2001.

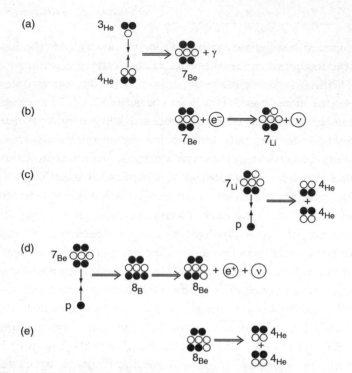

Figure 5 Making Helium, via Beryllium, Boron and neutrinos.
The process in fig 4 happens 85% of the time. Nearly all of the remaining 15% is due to helium-3 and helium-4 combining to make a single nucleus of beryllium-7 along with a photon (fig a). Beryllium-7 contains 4 protons and 3 neutrons. In fig (b), the beryllium-7 may pick up an electron, turning into lithium-7 (3 protons and 4 neutrons) and a neutrino (with an energy of less that 0.9 MeV). In (c), the lithium finally combines with a proton to make two nuclei of helium-4. There is a very small chance (about 0.01% or 1 in 10,000) that the beryllium-7 fuses with a proton to make boron-8 (5 protons and 3 neutrons) which then decays turning into beryllium-8, emitting a positron and a neutrino (fig d). This neutrino can have energy as high as 15 MeV and it is these relatively high energy neutrinos that Davis' experiment could detect. Finally, (fig e), the beryllium-8 then splits into two nuclei of helium-4. Protons are denoted by solid dark circles; neutrons by open white circles.

of city lights, so Davis's experiment would have to go deep underground to get away from the effects of the Northern Lights.

The Barberton limestone mine in Ohio is 700 metres deep. Inspired by the news that the production of beryllium-7 was large, at the end of 1959, Davis and a colleague, John Calvin, installed the tank from the Savannah experiment in the mine and started taking measurements. They were looking for neutrinos produced when beryllium-7 combined with a proton to make beryllium-8. They had only just begun when there was bad news: whereas nuclear physics experiments had shown that production of beryllium-7 was easy, the critical next step, where it combines with a proton to make beryllium-8, turned out to be difficult. The pessimism that this news generated was reinforced as, once again, Davis failed to see any convincing evidence for solar neutrinos.

Detecting neutrinos produced in nuclear reactors here on Earth had been hard enough; by 1960, it was becoming abundantly clear that looking for solar neutrinos was going to be a tough business at best, if not impossible. Fred Reines, who with Clyde Cowan had first detected the neutrinos produced by nuclear reactors, wrote a review article that year[x] in which he concluded that a search for solar neutrinos would probably be unsuccessful 'even with detectors of thousands or possibly hundreds of thousands of litres of C_2Cl_4,' (tetrachloroethylene or cleaning fluid) and that the uphill battle 'tends to dissuade experimentalists from making the attempt'.

By this stage, Davis was deep into his quest, in all senses of the word – 700 metres deep might not be deep enough, and 4000 litres was certainly too little. If Reines was right, Davis was chasing a phantom. It was during this depressing time that Willy Fowler was sent a paper to referee by a young theorist named John Bahcall. As a result of this chance encounter, Bahcall and

65

Davis would be thrown together in a quest that would absorb them for the rest of their lives.

Enter John Bahcall

John Bahcall never expected to become a scientist. Born in Louisiana in 1934 his main interest at school was playing tennis until in his senior year he discovered that he had an academia-related talent: debating. He and a colleague won the US national high school debating competition in 1952 and on the back of this he enrolled at the state university to study philosophy. His ambition at that time was to become a rabbi.

At the end of his first year, he enrolled in a summer course, at the University of California in Berkeley, and loved it. He then managed to register at Berkeley as a full-time student. His interest was still philosophy but to graduate, it was necessary to include a science course. It was through this that he fell in love with physics.

In his philosophical studies, he had read Bertrand Russell. It was Russell's remark about the insignificance of humans in the universe that inspired Bahcall's interest in astronomy. However, his career would only come to that field by a series of chances.

In 1960, he was at Indiana University attending lectures on the theory of the weak interaction[25] – the force that controls beta decay and the behaviour of neutrinos, the theory of which had begun with Fermi's ill-fated paper in 1934. In order to deepen his understanding of beta decay, Bahcall made up problems for himself to solve. These included the inverse of beta decay – where an electron

---

[25] In chapter 7 we will see that by 1960 the weak interaction had begun to take centre stage in physics.

would be captured along with the release of a neutrino. The standard lectures dealt with the capture of electrons from their orbits within atoms; as an exercise to check that he understood the principles, Bahcall calculated what would happen if the electrons were initially flowing freely instead of being trapped inside atoms.

One day, he was having lunch with an astronomer, Marshall Wrubel, who asked what he was doing. Bahcall told him about his calculations, but expressed disappointment that when he put numbers into the equations that he had derived, it didn't look as if anything he had calculated could ever be measured. That was the problem with neutrinos – theoretically fascinating, but experimentally their effects were at, or even just beyond, the limits of detectability.

Wrubel then made the remark that would determine the rest of Bahcall's life, that electrons being captured in free flow was what could happen inside stars. Perhaps there, Wrubel suggested, was where there might be some hope of bringing his ideas to fruition. The place to start, he added, was the classic paper by Margaret and Geoffrey Burbidge, Fred Hoyle and Willy Fowler, which explained how the elements are formed in stars, and was regarded as the bible for nuclear astrophysics.

When Bahcall read the paper, he found that, in a table at the back, Fowler had listed the characteristic properties of atomic nuclei that were involved in forming the heavier elements. Many of these involved beta decays, which were important as they were the slowest, and as such set the timescales for the light elements to build up to heavier ones. Fowler had assumed that the rates were the same as measured in the laboratory, but the calculations that Bahcall had done had shown him that the rates for electrons to be captured in free flow – as in stars – and from trapped orbits – as on Earth – were different. In short, one of the basic assumptions in this seminal paper was flawed.

The implication was that the chances for electrons being captured in the cool laboratory conditions on Earth and in the plasma of a star need not be the same. While the effects of the Sun's huge temperature could to some extent be simulated in the laboratory, by colliding the relevant atomic nuclei at energies corresponding to these temperatures, the Sun's huge central density, fourteen times that of lead, could not. However, quantum mechanics can be used to calculate its effects. That is what Bahcall had done.

He wrote a short paper pointing out that the rates for beta-decay processes in stars would differ from those being used by astrophysicists. He submitted it for publication to *Physical Review*, and it duly appeared.[xi] To his surprise, and before the paper had been published, he received a letter about it, from Willy Fowler himself. As Bahcall had not sent out any copies other than the one submitted to the editor, this could only mean that Fowler had been asked to referee it. It was no surprise that Fowler would be the natural choice: it was after all his table that had inspired Bahcall to look into this. The surprise was that the letter was offering him a job to work with Fowler at Caltech.

Fowler had acted decisively. In addition to attracting Bahcall to Caltech he had written another letter, this one to Ray Davis.

Recall that, in the quest for solar neutrinos, the excitement when it turned out that beryllium-7 was a thousand times easier to make than previously thought had been dampened by the experimental discovery that the critical next step, where a proton hit the beryllium-7 turning it into beryllium-8 and producing the crucial neutrino, would be difficult. However, in light of Bahcall's paper, there was potential doubt as to whether the experimental measurements involving beryllium and protons in the laboratory might necessarily imply such bad news for producing neutrinos in the Sun. However, Bahcall's calculations had not included this particular example.

When Fowler saw Bahcall's paper and realised its significance, he wrote to Davis that there was 'a guy in Indiana' who knew how to calculate how nuclear physics works in the Sun. And so it came to be that, in February 1962, Davis wrote a historic letter to Bahcall asking about this specific process. Bahcall started calculating.

By 1963, his first attempt was complete. It didn't give much encouragement. Bahcall's numbers showed that there was a difference between what had been measured on Earth and what should happen in the Sun, but even when this was taken into account it meant that a 4000 litre tank would only capture one neutrino every 100 days, fewer than a handful in a year. Nor did it encourage building a larger experiment, as even 400,000 litres would only capture one neutrino a day. By and large, astronomers were not interested in what was viewed as an expensive experiment anyway, let alone one that looked unlikely to be able actually to detect the solar neutrinos.

Davis was different and was eager to build a 400,000 litres experiment. First, his experience with the 4000 litre experiment in the Barberton mine made him confident that an increase by a factor of 100 was feasible. He also felt that a tank of this size could work efficiently: Davis was trained as a chemist and he was sure that he would be able to extract even the very few argon atoms that would be the 'smoking gun' for solar neutrinos. Furthermore, he believed that he could make the tank sufficiently leak-proof to avoid contamination by argon from the air and surroundings. This protection would be critical if he was to be sure that a mere handful of these atoms had indeed been produced inside his apparatus.

The major problem looked likely to be cosmic rays getting through to the experiment, producing argon when they collided, and being mistaken for neutrinos. To beat this, he and Bahcall concluded that the enterprise would need to be at least 1220 metres underground. Where were they to find a suitable cavern, deep enough and large enough? Even if there was one, would it be suitable in practice for a scientific experiment? As of 1963, the venture was regarded as a huge risk; few thought it likely to succeed.

Nonetheless Davis and a colleague, Blair Munhofen, started searching for deep mines in the United States. How did one go about this in the pre-Google™ age? The answer was to consult the national Bureau of Mines, who recommended two possibilities that appeared to meet their requirements: the Anaconda Copper Mine in Butte Montana, and the Homestake Gold Mine in Lead, South Dakota.

When Davis and Munhofen visited the mines to take a look for themselves, they found both good news and bad news. The owners of the Anaconda mine were keen for their site to be used and quoted a cheap price for providing a concrete lined cylindrical hole 1280 metres down. Unfortunately the cavern was too small. The Homestake mine looked more promising. Here, size was no problem. A cavity large enough to house a 400,000 litre detector, a volume the size an Olympic swimming pool, could be opened up 1480 metres underground. So far so good. However, the estimated costs of excavating this at Homestake were very large and so they decided to carry on searching for a site.

They came across Sunshine Mine in Kellogg, Idaho. This silver mine was 1640 metres deep, the rock was strong enough for excavation, and the costs at last seemed reasonable. Even though there was no money for their proposal, nor even any formal promise by any agency to fund it, at least they knew there was somewhere that a 400,000 litre experiment could be done.

UNDERGROUND SCIENCE

In the 1960s, the Neils Bohr Institute in Copenhagen was one of the leading centres for nuclear physics in the world. Among the faculty, Aage Bohr, son of Neils, and Ben Mottelson were at the height of their creative powers, building on their theory of nuclear structure that was to win them the Nobel Prize in 1975 (shared with the American, James Rainwater). It was in the summer of 1963 that Bahcall visited the Institute to give a talk about his calculations. What happened would change everything.

He began with a review of the nuclear physics involved in solar fusion and displayed his calculations of the numbers of neutrinos these should produce. The experts in the audience agreed with what they were hearing. Then he moved on to how these neutrinos were to be detected, describing how they would be absorbed by the chlorine which was then converted into argon. It was at this point that Ben Mottelson noticed something.

He realised that Bahcall had calculated the rate assuming that the neutrino converted the chlorine directly into argon. This could indeed happen, but what Mottelson had noticed was that the solar neutrinos had enough energy to make the argon nucleus with more internal energy than it normally has in its 'ground' state, sufficient to make an 'excited' state where a neutron in chlorine is just changed into a proton without the rearrangement needed to form the ground state. The excited state could then relax to its normal state, emitting the excess energy as a gamma ray. It seemed to Mottelson that this might actually be easier than the process that Bahcall had focused on.

'Have you looked at this?', Mottelson enquired. Bahcall admitted that the possibility had not occurred to him.

This was an intriguing idea. Bahcall set to work to see what effect it might have, and the answer turned out to be everything he had hoped for: it was 20 times easier for chlorine to capture neutrinos this way. The implication was that Davis's detector would capture neutrinos 20 times faster than previously thought. Whereas Bahcall's calculations had been predicting that Davis would capture merely one neutrino a week, now suddenly there was the possibility of a handful each day. Even though this was still a small number, it began to offer a tantalising hope of success. Optimism returned.

In November 1963 they presented their ideas on the feasibility of a 400,000 litre chlorine detector of solar neutrinos to an international conference on stellar evolution, at the Institute for Space Studies in New York. The reaction was so low-key that in the closing speech summarising the conference it was completely ignored. Undeterred, and sure that they had come up with a realistic experiment, they went to Brookhaven National Laboratory to convince its director, Maurice Goldhaber, to allocate some of the laboratory's science budget towards supporting the enterprise. To have any chance of a positive response, given the lack of enthusiasm that had been forthcoming from the astrophysics community, they decided to tailor their pitch to Goldhaber's interests.

Davis knew that Goldhaber, a distinguished nuclear physicist, was sceptical about astronomers 'being able to say anything correct about anything interesting'[xii] so there was little to be gained by emphasising the solar aspect of the experiment. Bahcall, however, was young, excited and 'full of calculations that I'd done about the Sun'. Davis explained to Bahcall that Goldhaber distrusted astrophysicists, and then demanded that they agree on tactics. Davis insisted that Bahcall trust him, restrict his remarks to the nuclear physics of the much increased capture rate and how this idea could be tested at Brookhaven, and let Davis talk about the experiment. Bahcall reluctantly agreed.

As Davis had hoped, Goldhaber was very interested in the nuclear physics ideas and, incidentally, also approved the solar neutrino experiment. Davis and Bahcall had, after all, dared to mention this motivation. What they had said was that if the experiment showed the rate of solar-neutrino capture to be different from what the theory predicted, it would confirm Goldhaber's conviction that astrophysicists 'did not really know what they were talking about'.

In order to further their case for the fully fledged experiment, they each wrote a paper: Davis on the proposed experiment and Bahcall on the theory behind it. These appeared back to back in *Physical Review Letters* on 16 March 1964. Davis's paper reported the results of the trials with the 4000 litre detector, consisting of two separate 500 gallon tanks located in the limestone mine, 700 metres below Barbeton Ohio. The care taken was impressive. As the signal for solar neutrinos would be at most a handful of argon atoms, and as air itself contains this element, they had initially purged the tanks with helium gas to remove every trace of it.

After the experiment had been running for 18 days they checked to see if there were any signs of radioactive argon. The good news was that they had some, but far too little to say for sure if these traces were caused by solar neutrinos, by other background activity or were even left over from the air when they originally purged the tanks. Nonetheless, the fact that they could measure such small amounts, which were just on the edge of discriminating between signal and background, showed that the idea was doable in principle. They calculated that 400,000 litres of fluid would be enough to improve the signal relative to the noise. However, to make the experiment work in practice, it would need to be much deeper, perhaps 1370 metres below the surface, in which case they estimated that nine out of every ten radioactive argon atoms would be due to solar neutrinos.

At least they knew of a suitable venue, the Sunshine mine. What they needed were the finances. They also needed to convince others that they could in fact achieve the task.

Bruno Pontecorvo held a special seminar in Leningrad to report on Davis and Bahcall's papers. There was a lot of interest but Pontecorvo later said that he was the only person present who believed that the experiment would be successful. There was wider publicity, not universally appreciated, courtesy of an article in *Time* magazine. Whereas science today has a high media profile, and scientists are ever ready to publicise their work, in 1964 this was regarded less favourably. However, the publicity in *Time* had unexpected benefits in helping to advertise their search for a suitable mine and in procuring a satisfactory tank for their detector. Davis would later say 'these tank people [took] us more seriously after the article in *Time*'.[xiii]

Goldhaber must have been convinced, because money for the experiment came from the chemistry budget at Brookhaven. No formal proposal was ever submitted to a federal funding agency.[xiv]

Work Begins

At last they had the funds, but suddenly there was no mine. Plans to build the experiment in the Sunshine mine fell through; the Homestake mine was available but too expensive. The publicity from *Time* now came to their aid. The management of Homestake mine was asked to reconsider the project, and it duly came up with a lower estimate: excavation could be done for $125,000 and work could start in the spring of 1965. There was also the added advantage of a bigger chamber than would have been the case in the Sunshine mine.

Excavation of the rock began in May 1965 and the cavern was ready by August. Davis and Blair Munhofen, his colleague who had done most of the negotiating with the mine, descended the shaft with their hosts, the darkness broken only by the miners' lamps on their safety helmets. They were guided into the cavern and started looking around with these lamps when suddenly the lights came on and illuminated the void. They gazed at the enormous room, ten metres across and twenty metres long with a ceiling ten metres above them; chain-link fencing on the walls, the floor concreted with pedestals to support the tank, and a monorail for the lifting hoist on the roof above them. The Homestake people and the scientists were all very

75

pleased. The challenge now was to build the tank and get it down the hole.

The Chicago Bridge and Iron Company (CBI) built the tank. On the scale of experiments that had been the norm in those days, the scientists thought this was a big affair. The CBI people by contrast found it very small, and later said that they would not normally have been interested in building what they regarded as a small conventional tank, but 'were intrigued by the aims of the project and the unusual location'.[xv] Another feature of the tank was that it had to be completely sealed to prevent any air, and hence argon, leaking in. Here CBI were masters: they had built space chambers for NASA. The vessel was completed in 1966, its inside thoroughly cleaned by sand blasting and scrubbed with solvent. They tested its radioactivity so as to estimate how many conversions to argon might occur due to the natural radioactivity of the apparatus. At last everything was ready for the tank to be filled.

The 400,000 litres of cleaning fluid had to be bought, brought to the site and taken a mile below ground. Ten railroad cars full of the stuff were brought from the Frontier Chemical Company in Wichita, Kansas, to the site in Dakota. The liquid was then put into specially designed tanks, each carrying 2500 litres and capable of fitting into the shaft, the hoist and the underground rail system of the mine. Loading, transporting and emptying each individual tank took several hours; the 150 trips took five weeks with the aid of the Homestake hoist-man and five scientists. Once this was completed, a whole series of purges were done to remove all traces of air, not just the air that had been in the tank already but air that had become dissolved in the chlorine itself.

By the end of the summer of 1966 the experiment was ready to begin. The total cost was $600,000, or as Davis described it when asked at a conference: 'Ten minutes of time on commercial television'. Twenty years had passed since Pontecorvo first suggested chlorine as a way of detecting neutrinos; seven years had already elapsed since Davis's first failed attempt. Little did anyone anticipate that another 30 years would pass before the full import of what they were about to do would be understood.

How many SNUs?

With the experiment in place and ready to start, the question was how many neutrinos were they expecting to find? Bahcall had steadily improved the precision of his calculations over the four years since he had become committed to the quest. These incorporated everything he knew about the workings of the Sun and the various nuclear reactions that were believed to power it. From all this, he computed the energy and the number of neutrinos that the Sun emits each second.

These neutrinos spread out over all space such that the Earth is permanently irradiated by this torrent. When he put all of this together he concluded that 66 billion solar neutrinos cross a square centimetre (about the size of your eye socket) each second. This is the total number but there are several different ways that they can be born. As a result, they don't all have the same energy. Critical for Davis's experiment would be how many of them would his detector, using chlorine, be sensitive to?

According to Bahcall's calculations, sixty of these billions originate in the primary fusion reactions whereby nuclei of hydrogen – protons – combine in a series of steps to make helium-4. The vast bulk of these individually would have too little energy to activate the chlorine in Davis's detector, and so

77

he had no chance of capturing them. However, neutrinos are produced in other processes because not all of the nuclear reactions end up as helium-4. As figure 5 on page 63 showed, it takes several steps to make helium-4 and, along the way, collisions can take place that produce different end-products. For example, at an intermediate stage, helium-3 has been made. However, as we have seen, it is possible that one nucleus of helium-3 may hit a nucleus of helium-4 that had been made earlier.

This fusion of helium-3 and helium-4 makes beryllium-7 and also produces a neutrino. Bahcall estimated that some 5 billion out of the 66 billion per second penetrating your eye are born this way. Nor is this the end of it. Half of the Sun still consists of free protons, and the beryllium-7 in turn might combine with one of these to make boron-8. Here again a neutrino is emitted, and what's more, with enough energy that were it to bump into an atom of chlorine in Davis's detector, it would be recorded.

That is the good news. Unfortunately this critical reaction is rare: as we have said already, out of every 10 billion solar neutrinos, a mere one million – one ten-thousandth of the total – come from this late stage. Davis's detector would therefore be blind to all but these relative few.

So at best Davis would be able to see but the faintest glimmer out of the whole spectrum of solar neutrinos. And to make matters worse, nearly all of these would pass through the whole Earth, let alone his detector, without disturbing anything. How many solar neutrinos could Davis hope to capture? Bahcall factored this into his formulae in order to come up with the final answer. He expressed the number in 'SNU', pronounced 'snew', which stands for solar-neutrino-unit. As this has become part of the standard lexicon of modern physics it is worth taking a moment to say what it means and why he invented it.

Starting from Fermi's theory of beta decay, it is possible to work out the chance of a neutrino hitting an individual atom and revealing itself. As we have said earlier, the number is so tiny that it had been thought to be as good as nothing. Bahcall used the theory[26] to compute that for a neutrino that had been born along with boron-8, the chance of it hitting a single atom of chlorine-37 was only one in 10 followed by 35 zeros per second, or 10^{-36}. Put another way, it means that an atom of chlorine-37 would have to wait 1 ... and 36 zeros ... seconds, that is some ten billion times longer than has elapsed since the Big Bang, before it had a 50:50 chance of capturing one of these neutrinos. It is obvious that some more friendly way of counting numbers was needed. Rather than saying a mouthful like 'one in 10 to the minus 36 per second' it has become traditional to refer to this capture rate as 1 SNU.

This is a small number indeed, but fortunately nature also gives us some big numbers that can enhance it. Bulk matter contains trillions upon trillions of atoms, each one of which has this tiny chance. A little multiplied by a lot may be measurable, and in 400,000 litres of cleaning fluid there are lots of atoms of chlorine-37: about 2 followed by 30 zeros, or 2×10^{30}, in all. So the average waiting time for a single capture by a random atom somewhere within the tank, if the capture rate is 1 SNU, would be only about half a million seconds, or six days. Thus if you take Bahcall's predicted number of SNUs it gives you the number of captures in six days.

---

[26] Fermi's theory had actually been refined following the discovery that mirror symmetry, 'parity', is not found in processes involving neutrinos; this is described in chapter 7. Bahcall correctly used this more sophisticated version of the theory in his calculations.

To compute the number, Bahcall used the best models of the solar interior, factored in data on various nuclear reactions that had been measured in experiments on Earth, and included the chlorine capture mechanism that Mottelson had suggested. Having put these all together he announced the answer: the rate would be 7.5 SNU with an uncertainty that meant it could be up to 3 SNU larger or smaller than this.[27] About 80% of this expected rate would be neutrinos produced in the decays of radioactive boron-8, which Bethe had long before predicted would be formed when beryllium-7 captured a proton from the Sun's primary fuel, and which Davis's experiment would hope to detect.

By 1968, two years after the experiment had begun, Davis was prepared to announce his first results: if he was seeing any solar neutrinos at all, the numbers were far too small. Assuming that the experiment was working properly, he was finding a value that was at most 3 SNU. If Bahcall's calculations also were right, there was a problem somewhere.

On the one hand, Davis's experiment was the only one to claim to have discovered solar neutrinos. This was a singular achievement, but it failed to make headlines as the shortfall made many worry about the reliability of the investigation as a whole. How sure could Davis be that what he was measuring were even solar neutrinos at all? How well sealed was his detector? Could argon get in from outside, or be produced some other way; could Davis convince sceptics that he could really measure such small numbers of atoms in such a vast assembly?

---

[27] This result turned out to be robust because in the subsequent 20 years, as Bahcall and others refined their calculations, the predictions never fell outside this range. By 1980, this had settled to the same value of 7.5 SNU but with the uncertainty much reduced to 1.5 SNU, see chapter 8.

Willy Fowler had challenged Davis about this: inject 500 atoms of radioactive argon-37 into the tank, stir it up and then extract them all. Davis did so, and extracted every one.

There were those who were convinced that the experiment was in fact doing what it claimed to do, but disagreed on what it all meant. Those who were not primarily astrophysicists decided that Davis's data showed that if the Sun indeed produced fewer neutrinos than the standard solar model predicted, then that was the end of the solar model. Many held the same opinion as Goldhaber, who had finally agreed to underwrite the experiment, that astrophysicists by and large did not know what they were talking about. The latter, for their part, insisted that they did, and that something else was wrong. Perhaps the data, which were not large in number, might be fooling us; toss a coin a few times and it might come up heads each time just by chance, but if it continues to do so after more trials, it probably means that you have a double-headed coin. To meet this challenge it would be necessary to improve the efficiency of the detector and to carry on collecting data.

The size of the apparatus was already fixed, and the recovery of the argon and chemical analysis was nearly perfect, so there was little room for manoeuvre there. The background from cosmic rays was known to be small, but the signal from solar neutrinos was also turning out to be small, so the relative effect of the background was actually big. The best chance for improving things would be to reduce the background noise further somehow. Even at a depth of a mile underground, cosmic rays would create two atoms of argon-37 in the tank each week, and for the experiment to be convincing, this would have to be reduced to at most one a month. That was the answer; the question though was how could it be done?

The 'Swimming Pool Improvement'

Here they were with a tantalising situation: they were potentially the first humans ever to look inside a star, but they were unable to convince anyone that they had succeeded. Their 'camera' was not quite good enough; somehow they had to kill the background so that the faint signal would stand out.

They brainstormed ideas at coffee breaks, over lunch and after work at the Caltech swimming pool. It was while relaxing there one afternoon that they had a stroke of luck. The astronomer, Gordon Garmine, saw them and came over to talk. He told them that he had heard about the chlorine experiment, and of their problems with the background, and wondered if a trick that was used in his field might help. This was based on something that electronics experts called 'pulse rise-time discrimination'. This sounds like something from a bureaucratic manual, but the words do describe exactly the phenomenon.

The electric pulses in certain types of detector take time to rise to a peak. When electrons are captured, as in the events that neutrinos induce, the pulse rises very rapidly to its maximum; this is quite different from the slowly rising pulses that background cosmic rays would cause. This sounded an ideal way to filter out the unwanted background events. But could it be done?

Davis consulted technicians at Brookhaven who said that the idea worked in many cases, where it differentiated fast from slow, but the particular circumstances that Davis was interested in were more tricky. In effect, the pulses from neutrino events would rise very fast while those from the background would be not quite so fast, in other words, not slow enough to be discriminated in practice. Or at least, not with the amplifiers then available.

Within a year, the Brookhaven electronics experts had developed devices fast enough to do the job. By 1970, with these new amplifiers installed, Davis was able to reduce the background events to one per month. Davis would always refer to this critical development as the 'swimming pool improvement'.

This led to a brief period of mild panic. In November 1971, Davis phoned Bahcall to tell him that not a single sign of a neutrino had been found for two months in the latest experiments including the new sensitive instruments, which left the awful possibility that the experiment was about to reveal that there are no neutrinos arriving at all! This turned out to be one of those statistical flukes, like a run of reds on the roulette wheel as you continue to place your money on black. Solar neutrinos did continue to arrive, though there were still too few of them.

There were improvements in the theory too. The various nuclear reactions that powered the Sun were being measured in laboratories more precisely than before, and using these better data as input to Bahcall's calculations, confidence in his predictions grew. In 1972, after Davis had accumulated four more years of data with a steadily improving detector, Bruno Pontecorvo wrote to Bahcall that 'It starts to be really interesting! It would be nice if all this ends with something unexpected from the point of view of [neutrinos]. Unfortunately it will not be easy to demonstrate this even if nature works that way.'

Very few people worked on solar neutrinos from 1968 to 1988. Davis's chlorine experiment was the only one recording data during those two decades. Bahcall pithily summed up the situation in a tribute to Ray Davis: 'All the people working steadily on solar neutrinos, theorists and experimentalists, could (and often did) fit comfortably into the front seat of Ray Davis's car'.[xvi]

Bahcall had updated and refined his calculations, and stood by his numbers. Davis continued to improve his experiment. By 1978, a decade of disagreement between the results of Davis's experiment and the predictions of the standard solar model had led to an impasse. A conference of scientists got together at Brookhaven that year to decide what to do next. Clearly a new experiment was needed, and ideally one that would detect neutrinos from the primary proton–proton fusion process directly, rather than the relatively minor numbers originating from boron-8 that Davis had been detecting.

For physicists at large, the situation was known as the solar neutrino problem, and metaphorically put in a drawer awaiting someone to find out the source of the perceived error so that the saga could be dumped in the wastebasket of wrong science. However, as a growing number of scientists continued to examine Bahcall and Davis's work, the consensus strengthened that the apparatus appeared to work as it should, and that the calculations were robust.

If there was one area of doubt, it was with the extreme sensitivity of Bahcall's numbers to the assumed temperature at the centre of the Sun. The value was proportional to this temperature multiplied by itself over and over 25 times. A 1 per cent change in temperature would lead to a 30 per cent change in the number. A 10 per cent reduction in temperature could explain the shortfall, and it was this that gave rise to the 'Is the Sun still shining?' mantra. As we saw on page 44, many novel ideas suggested that maybe the Sun has stopped shining and that a real energy crisis is on its way.[xvii]

Relatively little attention was paid to the possibility that the Sun might be innocent and that the neutrinos were to blame. Could the neutrinos have disappeared en route? One wild idea was that there are higher dimensions than the three space dimensions that we perceive such that in their 150 million kilometre journey neutrinos have a chance to escape into some sort of parallel universe, and disappear from our view. The arguments went on for 20 years. Ironic then that Bruno Pontecorvo had proposed the solution in 1968 as soon as Davis's first results appeared. He was not primarily interested in the Sun. Instead he focused on the neutrinos. Soon after Cowan and Reines had proved that the neutrino existed, others had started noticing unusual things about it. It was these that Pontecorvo would seize on in his next seminal contribution to the neutrino story.

7

ONE, TWO, THREE

While Ray Davis was having a hard time finding neutrinos from the Sun, experiments with terrestrial ones had turned out to be remarkably successful. In Fermi's theory, neutrinos aren't always shy. Neutrinos passing through the entire Earth 'as easily as a bullet through a bank of fog' is true when they have only small amounts of energy, such as when produced in beta decay or in the Sun, but if they have very high energies, they are much more likely to bump into matter and be revealed. As a result of this insight, neutrinos were about to become part of mainstream science.

The idea that the chance of a neutrino interacting grows with energy was implicit in Bethe and Peierls' 1934 paper, and its implications were articulated by Mel Schwarz in the USA in 1960. This was one year after a seminal paper by Pontecorvo, which also had realised the potential for using beams of high energy neutrinos. While important, this was not the main reason why Pontecorvo's paper became so famous. It contained another insight, so profound that it would change the perception of neutrinos and lead to the modern 'standard model' of particles and forces. He had stumbled on a deep and far reaching truth: not all neutrinos are equal; some are more equal than others.

As we have seen already, Pontecorvo was by then ensconced in the Soviet Union. As a result his paper appeared in a Soviet journal and, having been written in Russian, it remained unknown in the West until its English translation appeared. We will come to the implications of that later. First though, let's see how things stood by 1959, setting the scene for Pontecorvo's triumphs, and tragedy.

Who Ordered That?

By the 1940s, the fundamental building blocks of atoms had been identified as electron, proton and neutron. The neutrino, still years from direct observation, was likened to an electron but without electric charge or mass, and of dubious identity. Overall, matter seemed well-ordered, constructed from these few particles, though the laws governing their behaviours were still being worked out. Then, out of the blue – in a sense literally – cosmic radiation revealed new particles.

One of these was the pion, π. This at least had been predicted as a solution to the paradox that atomic nuclei exist, even though their constituent protons repel one another electrically. The cause of nuclear stability is that when nucleons (neutrons and protons) are very close, they feel a strong attractive force, which overpowers any electrical repulsion. According to quantum theory, when protons or neutrons collide at high energies, one effect

87

of this nuclear force is to convert some of their energy into an ephemeral particle – the pion, π.

So far, so good. However, the cosmic rays also revealed that there is another member in nature's players. This is the muon, μ, which appeared to be nothing more than a heavier version of an electron – an insight that Pontecorvo was one of the first to articulate.[xviii] 'Who ordered that?' physicist Isadore Rabi famously exclaimed, and more than 30 years were to pass before even the beginnings of an answer emerged. The muon had no obvious place in the elements of matter known on Earth. Both the muon and the pion produce neutrinos when they decay, though this was not known until years after their discovery.

It is ironic that the muon was discovered first (in 1937) and originally thought to be the anticipated pion. In fact, it is the progeny of a pion.

The first hint that the particle was not the pion was that it showed no strong affinity for linking to the atomic nucleus. As the whole rationale behind the prediction of the pion had been that it was the agent that binds the nuclei of atoms, the apparent lack of interest by the candidate particle made no sense – unless it was not the pion after all. The possibility that there were in fact two particles – the one that had already been discovered being what we now call the muon, and the nuclear pion still to be found – had been suggested by theorists in Japan and also the USA in the period leading up to the pion's eventual discovery in 1947. That same year, Pontecorvo produced compelling arguments that the muon was not the nuclear agent, but instead was like a heavy version of the electron.

He did so by showing that its behaviour in and around atoms and nuclei was more like that of an electron than something nuclear. He proposed that the beast was produced in reactions analogous to beta decay. Here were the first hints that the muon is like a heavy electron and not the anticipated nuclear pion.

Today, we know that when cosmic rays smash into atoms in the upper atmosphere, a torrent of pions is produced. Indeed, it had been by studying cosmic rays that Cecil Powell of Bristol University discovered the pion in February 1947. In the debris that comes from these collisions, the pion is the primary particle (it was this 'primary' aspect that led to it being named for π, the Greek symbol for 'p') and it was its decay in about one hundredth of a microsecond that produces the muon.

The force that destroys the pion and causes it to decay has become known as the 'weak' force. Its naming reflects that it is weaker than both the electromagnetic force, which holds electrons in their atomic terpsichore, and the strong force, which grips protons and neutrons together in the atomic nucleus. Other than the fourth fundamental force, gravity, the weak is the feeblest of all.[28] It is also the hardest to study.

The decay of an electrically charged pion most often produces a neutrino together with a muon carrying the same electric charge as the parent pion. The muon is also unstable and decays to an electron[29] and energy that is carried away by neutral radiation.

The pion and muon were relatively easy to make and study in the laboratory, and it was soon found that whereas the muon has an intrinsic rotational motion or 'spin' of one-half like an electron or a neutrino, the pion has none. This also fits with the idea that when a pion decays, the half-integer spin of the muon is

---

[28] The force between two particles due to gravity is very feeble. It is the fact that the gravitational force between all the atoms in a large body add together that makes it dominate in bulk matter such as falling apples and planetary orbits.

[29] If negatively charged, the positively charged muon produces a positron.

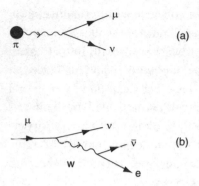

Figure 6 Neutrinos produced in the decay of pions and muons. (a) A pion decays into a muon and a single neutrino. (b) When a muon (denoted μ) decays into an electron, two neutrinos are required (actually a neutrino – v – and an antineutrino – $\bar{v}$)

counterbalanced by a counter-rotation of a half-integer spin neutrino. In the case of the muon decay, however, there is no change in the spin: the muon has one-half, as does the electron that it changes into. To all appearances, the muon looked like a heavier version of the electron, which has simply radiated off some energy and collapsed back into an electron. If that was the whole story the decay of a muon should produce an electron and a photon of light: $\mu \rightarrow e + \gamma$. However, a young physicist by the name of Jack Steinberger was about to show that this is not the case, and in so doing, uncover a great mystery.

Jack Steinberger

Jack Steinberger was born in Bad Kissingen (Franconia) in 1921. His father, Ludwig, was one of eight[30] children of a rural cattle dealer, who was religious teacher for the little Jewish community.

[30] J Steinberger personal communication. This differs from his autobiography on the Nobel website which attests to 12: http://nobelprize.org/nobel_prizes/physics/laureates/1988/steinberger-autobio.html accessed 28.3.2010.

His mother had the benefit of a college education, which was unusual for the time, and she supplemented the meagre income by giving English and French lessons, mostly to the tourists who provided the economy of the spa; Germany was living through the post-war depression.

After the Nazis had taken power, American Jewish charities offered to find homes for 300 German refugee children. Steinberger's father successfully applied on behalf of Jack and his elder brother. They arrived in New York at Christmas 1934. The owner of a grain brokerage house on the Chicago Board of Trade took Jack into his house, parented his high-school education, and also made it possible for Jack's parents and younger brother to come in 1938 and escape the Holocaust.

He studied chemical engineering for two years at the Armour Institute of Technology (now the Illinois Institute of Technology), but these were the hard times of the depression, and his scholarship came to an end. It was necessary to work to supplement the family income, which meant that he could only study chemistry at the University of Chicago in the evenings. The following year, with the help of a scholarship from the University, he was again able to attend day classes, and in 1942 finished an undergraduate degree in chemistry.

War having begun, he joined the army, where he experienced his first intense introduction to physics. After a few months studying electromagnetic wave theory in a special course given for army personnel at the University of Chicago, he was sent to the MIT radiation laboratory. The radiation laboratory was engaged in the development of radar bomb sights; he was assigned to the antenna group. His two years there offered him the opportunity to take some basic courses in physics.

After the Japanese surrender, he continued his studies at the University of Chicago. Here he was inspired by the courses of Enrico Fermi, which he recalled were 'gems of simplicity and clarity'. Fellow students included future Nobel Laureates C N Yang and T D Lee, as well as a roll-call of other physicists who made remarkable contributions to the field. 'There was a marvellous collaboration, and I feel I learned as much from [them] as from the professors', Steinberger recalled at his own Nobel address.

He wanted to do a thesis in theoretical physics, and Fermi took him on. This was during the interregnum where the pion had been predicted but not yet found; the muon had been found, but not predicted; and the belief that the muon was in fact the pion was beginning to fall apart. Pontecorvo had just proposed that the particle in question, which we now call the muon, appeared to be a more massive version of the electron – in effect, an electron containing much more energy. Experiments had shown that a muon decays into an electron, shedding energy in some invisible form. If it were true that a muon is simply a heavy electron, then it should have been able to shed this energy in the form of light, a gamma ray. However, this seemed not to take place; if it had, experiments studying what happened when muons stopped in matter and decayed should have found four or five times more electrons than were actually being seen. Something was wrong.

Fermi told Steinberger about this and suggested that the apparent shortfall could happen if the electron sometimes carried off less energy than the experiment had been sensitive to. In such a case the decay would not be recorded in the accounts. If a muon decays to an electron and a single photon, the energy of that electron is fixed. Steinberger and Fermi realised that if the electron were accompanied by two particles, making three in all, the electron's energy could take on a range of values, some of which would have been missed in the original experiment. Better still, Steinberger suggested a way to test this idea experimentally.

Fermi was intrigued, and as no one seemed prepared to try the experiment, he suggested that Steinberger do it himself. So he did. It took him less than a year from conception to its conclusion in the summer of 1948. And it confirmed his conjecture: the muon decays into an electron accompanied by two further particles, not one.

This showed that when a muon decays, whatever it decays to is not simply an electron and a photon.[31] The two 'missing' particles must have no electric charge, very little if any mass, and their combined spins cancel to nothing. They could have been photons, but if so, there was no reason why two were needed when one would have been enough and even easier. Some law had to be preventing a muon turning into an electron merely by shedding energy in the form of photons; the invisible radiation had to be something other than photons. Everything was consistent with the idea that it consisted of two neutrinos, which were escaping detection.

In a way, this is a replay of Pauli's original suggestion of the neutrino, as an agent for taking off 'missing' energy in beta decay. To modern ears, the suggestion that two neutrinos are being produced in muon decay might sound obvious; in 1948 when actual proof of the reality of the neutrino was still eight years in the future, it was far from obvious.

In 1956, the theoreticians T D Lee and C N Yang pointed out that processes controlled by the weak force might be distinct

---

[31] Or if it does, then so rarely as to be unobservable. Even today, in over one hundred billion muon decays that have been recorded, not one example of $\mu \to e + \gamma$ has been seen.

from their mirror image – a property that became known as parity violation. This was quickly confirmed by experiments on beta decay (of cobalt nuclei) and in muon decay, leading to Lee and Yang's Nobel Prize in 1957. The decays of charged pions also were caused by the weak force. If parity violation were present here, then for every 9999 pions that decayed into a muon and neutrino, one should decay into an electron (or positron) and neutrino, in effect, like 'traditional' beta decay. In 1958 Steinberger and some collaborators showed that when positively charged pions from an accelerator were stopped by the protons in a tank of liquid hydrogen, the resulting trails of bubbles could be used to distinguish decays into muons from those into positrons. They found over fifty thousand examples of decays into muons, and six clear examples of positrons – a ratio consistent with that expected if parity was being violated.

So the menu of 'beta' decays was growing. The traditional decays of nuclear particles produced electrons (or positrons) and neutrinos. Once in ten thousand times, the pion did also, but most of the time it decayed into a muon and a neutrino. Bruno Pontecorvo started wondering: are the neutrinos produced when a pion decays into a muon, the same as those emitted in conventional beta decays? With uncanny prescience, Pontecorvo once more was asking the right question.

A Tale of Two Neutrinos

By 1958, Fermi's theory of beta decay was established, up to a point. The neutrino had been discovered, and the rate that Cowan and Reines detected neutrinos in their experiment agreed pretty well with what Fermi had expected. The phenomenon of parity violation required some mathematical details to be revised, but the basic ideas remained. Fermi's theory implied that the chance of neutrinos reacting grew with energy. That would be good news for experiments at the new accelerators, but also had some absurd implications. If you imagined doing experiments at exceedingly high energies, beyond the ability of technology in 1958 but possible in principle one day, the theory implied that things could happen with probability greater than 100%.

This was illogical. The solution was to give up Fermi's idea that the particles involved all met at a single point. Electromagnetic forces were transmitted by agents, photons, so the idea that perhaps the weak force also has an agent, a W (for 'weak') boson, began to take hold.[32] While this would avoid the nonsense, G Feinberg pointed out that it had an implication for the decay of the muon. A muon decays into an electron and two neutrinos by the intermediate action of a W boson; the laws of quantum mechanics imply that you can do away with the neutrinos and have the muon decay into an electron and a photon, about once in every ten thousand times. However, by that time over a hundred million examples of muon decays had been recorded and not one involved this mode of an electron and a photon.

Feinberg did point out, however, that his argument assumed that the neutrino associated with the muon and that paired with the electron (Figure 7) were the same. Here was the first suggestion that the two neutrinos might actually differ. It was Pontecorvo who independently articulated this in its finest form, assessing all the evidence, suggesting experimental tests and

---

[32] This was confirmed in 1983 when the W boson was discovered at CERN. The story is told in *The Particle Odyssey*, Frank Close, Michael Marten and Christine Sutton, OUP 2002, and *Particle Physics: A Very Short Introduction* Frank Close, OUP 2004.

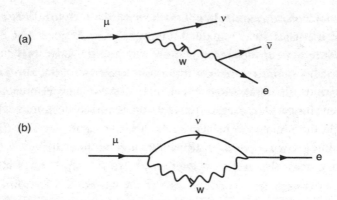

Figure 7 The W boson and muon decays. The W boson is the intermediate aagent in the decay of the muon, μ, into an electron and a pair of neutrinos – v and $\bar{v}$ – (shown in fig (a)). If these two neutrinos had the same flavour, it would be possible for a muon to convert into an electron and a photon courtesy of the W and neutrino momentarily occurring in an intermediate step, shown in (b)

pointing out that with the increasing energies of accelerators, such experiments, using beams of neutrinos, might become feasible in the near future. As we have already remarked, being based in Moscow, he wrote the paper in Russian and published it in the *Soviet Journal of Physics* in 1959; his work was unknown in the West until its English translation appeared in 1960.

The fact that the muon does not decay to an electron and a photon fascinated him. Forget the issue of whether it happens once in ten thousand times, for him the paradox was much starker: why did it not happen nearly all the time? Shedding a photon ought to be much easier than being destroyed by the weak force. If it didn't happen, he realised, something had to be forbidding it. Having been the first to articulate that the muon was a relative of the electron, Pontecorvo was now the first to propose that the muon is more than just a heavy electron: it has some special 'muon-ness' about it. Today, we call it 'flavour' (though precisely what this is remains a mystery).

Pontecorvo now took this idea and explored its wider implications: if a muon and an electron differ in flavour, why not the neutrinos also? He noticed that this makes a nice symmetry among the particles: an electron and a neutrino sibling is one pair, while the muon and its sibling neutrino form another. These became known as 'electron-neutrinos' and 'muon-neutrinos' respectively. In the shorthand notation of particle physicists they are written 'v_e' and 'v_μ', those symbols being the Greek 'nu', not just a curly 'v'.

He first showed that if 'electron-ness' and 'muon-ness' are preserved in nature, this would forbid the decay of a muon into an electron and a photon. He then listed ways of testing the idea. The most dramatic was that a neutrino, in addition to carrying energy, somehow also carries a memory of its provenance. Consequently, if a neutrino is produced along with an electron (v_e) or a muon (v_μ), then when it subsequently hits matter and picks up electric charge, its energy should materialise as an electron or a muon respectively.

This idea was clear, but as the most singular property of neutrinos had been their reluctance to show themselves at all, any hope of making fine detail measurements about whether they turned into electrons or muons appeared to be far beyond what could be achieved in practice. This is where Pontecorvo made his second important remark: the rates for neutrinos to interact rise with increasing energy, which meant that they could be become visible so long as you could somehow obtain high energy neutrinos. This led him to his insight: he suggested that the new high energy accelerators would be the place to do the experiment.

His idea was to make large numbers of high energy pions by first smashing a beam of high energy protons into a target. The pions decay into muons and neutrinos, which fly onwards in the direction that the original beam had been taking. A steel shield will absorb the muons, but would be almost transparent to the neutrinos. Several metres further on, you need another large target as a detector. The neutrinos will have high energy, and hence there would be a reasonable chance that occasionally a neutrino would hit an atom in the detector, pick up electric charge, and reveal itself. If all neutrinos are alike, the numbers of electrons and muons that are produced will be similar. However, if only muons appear, then the neutrinos carry an identity: electron-neutrinos differ from muon-neutrinos.

Being in the Soviet Union, he missed out on the experimental possibilities, because their facilities at that time were inferior to those in the West. Around 1960, this would become possible at Brookhaven in the USA and at CERN in Geneva, but not at Dubna, the laboratory near Moscow. He was not permitted to cross the Iron Curtain until the 1980s. Others would gain this prize.

Leon Lederman and Mel Schwartz

In the physics department at Columbia University in the 1950s, Fridays were traditionally the day for Chinese lunch, during which the latest problems in physics would be discussed. On Friday 4 February 1957, T D Lee first ordered the menu, and then announced that experiments using cobalt nuclei were beginning to look as if parity was violated in weak interaction. It was during the next-to-last course of the meal that an idea started to form in Leon Lederman's mind.[xix]

Lederman had been born and raised in New York, and was now on the faculty at Columbia. He had been doing experiments with pions and muons at the Nevis Laboratory, nearby in Irvington-on-Hudson, in particular measuring the decay of the charged pion into a muon and a neutrino. This was caused by the weak interaction, and he suddenly realised that here was a much easier way of looking for parity violation than the ongoing, slow, nuclear beta-decay.

At Nevis, a beam of protons hit a target and produced intense beams of pions. As the pions streamed out across the hall, about 20% of them would decay into a muon and a neutrino, which themselves carried on in the same direction that their parents had been moving. The pion has no spin, and so the neutrino and muon must spin in opposite senses to make the rotary angular momentum accounts balance. But if parity is violated, the massless neutrino only corkscrews in one sense (traditionally known as left-handed), which would constrain the muon's spin also to be restricted to just one orientation. The question was, how to measure the direction of this spinning muon?

The gift of nature was that the muon also decays by the weak interaction, producing an electron and two neutrinos. All that would be needed would be to stop the muons and see in which directions the electrons emerged. If parity was violated, the tendency for the muon to spin in just one sense would lead to more electrons emerging in the direction that the muons had previously been travelling – the forward direction – than in the opposite, backward direction. Immediately the afternoon's work in the university was over, Lederman rushed to Nevis Laboratory, where he told his student to modify the experiment that they had been doing, so as to check out his idea. During a quick dinner he spoke on the phone with a colleague, Dick Garwin, and invited him to join in and help. By 8 p.m. they were hard at work re-designing the experiment. Soon after midnight they started taking measurements, and some tantalising hints began to emerge suggesting that they were on the right track. Unfortunately, the accelerator was already promised for other experiments, and so they had to pause, spending the next day testing and improving the apparatus. On the Monday they were ready to proceed, but the accelerator maintenance teams had problems. It was not until Monday evening that they were eventually able to start in earnest. Garwin took the night shift. At 3 a.m. on the Tuesday Lederman was woken by a phone call: 'You'd better come in. We've done it.' By 6 a.m. there was no doubt; they were seeing a huge effect – more than twice as many electrons went forwards as backwards. Parity was not simply violated, it was utterly destroyed. Now it was Lederman's turn to wake someone up – T D Lee. By 7 a.m. they were getting calls from Columbia colleagues who were hearing the news, and by the end of the day, physicists around the world, in Geneva and Moscow, were repeating the experiment and verifying it for themselves. Lederman announced the results to an audience of 2000 at the annual meeting of the American Physical Society, in New York, on 6 February 1957.

As we saw earlier, the following year Steinberger measured the rare decay of a pion into a neutrino and an electron: 'normal' beta decay. There was by now no doubt that these decays involved the weak interaction, and that the missing ghostly particles were neutrinos. Lederman and Steinberger were soon to join with Mel Schwartz, a former student of Steinberger and by then an assistant professor at Columbia, in the collaboration that was to rewrite the neutrino lexicon. The story here also begins at one of the famous Columbia lunches. In November 1959, T D Lee was worrying about the paradoxical implications of Fermi's theory for the behaviour of weak interactions at high energies, and was leading a discussion on how to test it experimentally. It would be hard to study because when particles collide at high energies, the effects of electromagnetic and strong forces tend to obscure those of the weak force.

Mel Schwartz later recalled that 'lying in bed that night it came to me. It was incredibly simple. All one had to do was to use neutrinos'.[xx] The idea was that the production of pions, followed by their decays, might produce neutrinos in sufficient numbers that they could be used in experiments.

He wrote a short paper outlining the ideas, which was published in 1960. It was only then that Pontecorvo's paper appeared in English translation. At the end of Schwartz's paper he noted the 'related paper which has just appeared' written by Pontecorvo, and thanked Lee and Yang for emphasising the importance of high energy neutrino interactions.

The 'related paper' referred to Pontecorvo's mention of high energy neutrinos being the way forward. Pontecorvo's more profound ideas about the two distinct 'flavours' of neutrino were not in Schwartz's paper. However, in the meantime, Lee and Yang had also been thinking about what might be learned from these experiments. By the summer of 1960 they had reached the same conclusion that, unknown to them, Pontecorvo had already drawn: the absence of muon decay into electron and photon could be the smoking gun proving that the muon-neutrino and electron-neutrino differed. This became the quarry to chase.

The question though was how to do it? Although no existing accelerator was powerful enough to produce a neutrino beam with sufficient intensity to perform the experiments, at Brookhaven the new AGS (Alternating Gradient Synchrotron) was nearing completion. Leon Lederman calculated that the experiment could be done there and convinced Schwartz that this could really work.[33] A team of seven set to work: Schwartz, Steinberger, Lederman and four students and postdocs.

They used the accelerator at Brookhaven and fired its intense beams of protons into targets of beryllium. This produced large numbers of pions, which rapidly decayed into muons and neutrinos. A 13-metre-thick barrier of steel, built from the plates of an old battleship, filtered out the muons. The neutrinos passed clean through, and then downstream met 10 tonnes of aluminium. Over a period of ten days, more than 100 million million neutrinos passed through, out of which a mere 51 neutrinos gave themselves away when they hit the aluminium and picked up electric charge. However, every one of these 51 collisions resulted in a muon; none gave an electron. The trio of scientists had proved that muon-neutrinos and electron-neutrinos have distinct identities.[xxi]

Three decades later, in 1988, Lederman, Steinberger and Schwartz shared the Nobel Prize in physics. Their work had established that high energy beams of neutrinos can be made

[33] Memories differ on this. Schwartz in his Nobel address writes somewhat ambiguously: 'That evening the key notion came to me – perhaps [sufficient numbers of] neutrinos could be produced [for use] in an experiment. A quick back-of-the-envelope calculation indicated the feasibility of doing this at one or other of the accelerators under construction or being planned at that time. I called T D Lee at home with the news.... The next day planning for the experiment began in earnest'. Lederman in *The God Particle*, p. 290, writes 'Schwartz had somehow convinced himself that no existing accelerator was powerful enough to make a sufficiently intense neutrino beam, but I disagreed...I did the numbers and convinced myself and then Schwartz that the experiment was, in fact, doable'.

and used as a probe of the weak interactions, and during the intervening years they had variously pursued the weak interactions with these techniques. Their discovery that electron-neutrinos and muon-neutrinos are distinct became a foundation for the modern standard model of particle physics.

It is possible that Schwartz may have missed out on a second Nobel Prize. In 1971, he was using a similar experimental setup at Stanford Linear Accelerator Center (SLAC) in California. He found five unusual examples where neutrinos appeared to have interacted with the target but where no muon resulted. With hindsight, it is likely that he had made the first observation of 'neutral current' interactions – where a neutrino bounces off the target without picking up electric charge.

The SLAC management in 1971 had other priorities. Experiments there had just found evidence that protons and neutrons are not point particles but are built of more fundamental constituents, later called quarks. As this was looking very likely, but was not yet fully understood, that was where SLAC's primary effort was being placed. This also led to Nobel Prizes. Schwartz desperately tried to get funding for his own experiment from the National Science Foundation, but without success. I was at SLAC at the time and recall Schwartz repeatedly insisting that he had uncovered something utterly novel, and being most frustrated at his inability to follow it up. Soon afterwards he left particle physics and founded Digital Pathways, a computer company.

The 'neutral currents' were eventually discovered in 1973 by the 'Gargamelle' collaboration at CERN. This confirmed the theories uniting the electromagnetic and weak forces, which had in part been inspired by the discovery of the two neutrinos. In his Nobel address, Schwartz graciously noted that Pontecorvo had independently come up with many of the same ideas as had Schwartz, Steinberger and Lederman, saying 'His overall contribution to the field of neutrino physics was certainly major.'[xxii]

Postscript: Three Neutrinos

In 1976, a yet heavier form of electron, known as tau, τ, was discovered. The 'standard model' of particle physics had by then emerged, which predicts that every variety of charged lepton (electron-like particle) has a neutral counterpart (neutrino). So a third type of neutrino, associated with tau, was required to complete the tale.

If finding the v_e and v_μ had been hard, detecting the v_τ was even more so. The tau-neutrino will pick up charge and turn into a tau particle but the latter is very massive, more than twice as heavy as a proton, and so requires high energy neutrinos to begin with. The tau is unstable, and decays in less than a billionth of a second into a muon or electron accompanied by further particles. All this makes it hard to identify, even when it is present. Indirect evidence for its existence was found at LEP, the large electron–positron collider, at CERN in the 1990s, but direct observation was only achieved in 2000 by the DONUT experiment (Direct Observation of 'Nu-Tau') at Fermilab, Chicago.

The discovery of the tau particle in 1976 had been utterly unexpected, and won the Nobel Prize for Martin Perl. It pointed the way to the neutrino partner, so although the latter was very hard to find, its final observation was anticipated. Probably the most profound implication is that 30 years after the discovery of the tau, here was a further vindication of Pontecorvo's 1959 paper that identified 'flavour' as special, and that neutrinos carry identity cards.

The experiments at LEP showed also that three is the sum total of varieties of these light neutrinos. This implies that the electron, muon and tau are the extent of their electrically charged partners. In the standard model of particle physics, the proton, neutron and pion are made of 'quarks', and the six leptons are partnered by six varieties (flavours) of quark. These too have all been identified in experiments spanning half a century. So the humble arcane neutrino, with its three varieties, has enabled us to put a limit on the varieties of matter that nature uses. This is the first time in history that a limit to the number of fundamental particles has been found.

Why are there three flavours of neutrino? No one yet knows. However, in another of his seminal insights, Pontecorvo was to show that the fact that there is more than one plays a major part in the solar neutrino mystery.

8

MORE MISSING NEUTRINOS

All the Gallium in the World

By 1978, the data from Davis's experiment in the Homestake mine were getting better and more precise. The conflict between what was being measured and what was expected was deepening.

The rate of solar neutrino production in the rare process that he could detect was settling on 2.2 SNU, with an uncertainty of 0.4 SNU either way: it could be as large as 2.6 or as small as 1.8 by chance, or outside this range but with much less likelihood. Bahcall's theoretical computations for this number had also sharpened, and by 1980 his prediction stood at 7.5 SNU, with an uncertainty of 1.5. The basic theory had not really changed; rather it was the improved data on the fundamental nuclear processes that had increased Bahcall's confidence in his numbers.

Even if you interpreted the Homestake measurement to be as high as 2.6 SNU, and believed Bahcall's estimate of what they should have been finding to have been as low as 6 SNU, it was hard to avoid the conclusion that Davis was detecting less than half the number of neutrinos that he expected.

Nagging doubts remained as to how significant this discrepancy really was. The reason for such uncertainty was that Davis was looking only at neutrinos produced in a rare reaction that occurs as the end product in merely one or two out of every 10,000 initial *pp* fusions. The total rate that these particular neutrinos are produced depends sensitively on the individual rates for the various nuclear processes involved in the *pp* cycle. Although data on these had improved markedly as a result of the stimulus set by Davis's quest, and had in turn enabled Bahcall to sharpen his predictions, there was a clear need for a new experiment to find out if the problem lay with the theory of the Sun, or with the neutrinos.

This became one of the talking points at a conference about neutrinos that was held in Hungary in 1974. Bruno Pontecorvo

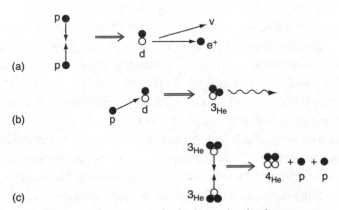

Figure 8 The *pp* proton-proton fusion reaction in the sun.

announced that he and three colleagues had plans for a 4km tunnel to be dug under the Caucasus, to site a dedicated neutrino laboratory. This would include a chlorine detector with almost two million litres of liquid, some five times that at Homestake, which hopefully could both confirm what Davis had done, and improve the sensitivity.

Useful though this might be, it would not get to the heart of the matter, which would require detection of neutrinos produced by the basic and dominant *pp* fusion process. The problem was that the energies of neutrinos coming from this primary reaction are too small to make chlorine react. As early as 1964, and with Davis's first failed attempt in his mind, Bahcall had remarked[xxiii] that if Davis's new experiment failed to see neutrinos, then a dedicated search for the lower energy neutrinos coming from the primary *pp* chain would be merited. Davis did see neutrinos, but never enough.

An advantage of such data, from a theoretical perspective, was that it was possible to predict the number of these primary neutrinos just from the visible luminosity of the Sun; there was no need for the detailed knowledge of nuclear reaction rates that had been the scourge of the work so far. That's the good news. The problem was that to detect these lower energy neutrinos, an experiment using gallium rather than chlorine would be needed. A successful experiment using gallium would be hugely expensive, not least because, to be practical, it would need about three times the annual world production of the element.

Davis and Bahcall had already devoted the better part of two decades to chasing solar neutrinos, so they were prepared to do whatever it took to resolve the issue. If this required them to build a gallium detector containing all the gallium in the world, at least they had to try.

To have any chance they needed to gather support. A paper in *Physical Review Letters*, co-signed by several experimental colleagues, described the advantages of a gallium detector, arguing that it would be sensitive to the neutrinos from the fundamental *pp* reaction, and would be able to distinguish among the various explanations of the solar neutrino problem. If they could build a gallium detector, the sensitivity compared to what they had done so far promised to be remarkable. Whereas Bahcall had calculated that for Davis's detector the capture rate of solar neutrinos would amount to 8 SNU, of which 1.2 came from beryllium-7, 6.2 from boron-8 and the remnant from his 'smorgasbord' of other nuclear processes, the gallium detector could be expected to be chasing a huge 132 SNU. The primary reason is that the gallium detector is sensitive to neutrinos of much lower energy, which chlorine does not capture. These include the all-important dominant ones from the basic *pp* fusion process, which on their own were expected to provide 74 SNU out of the total and were the primary reason for using gallium, but it also elevated the rate for the neutrinos produced by the other stages in the chain of solar reactions. Where the sensitivity to neutrinos from beryllium-7 and boron-8 had been a mere 7.5 SNU in Davis's detector, Bahcall estimated that a gallium experiment would push that rate up to 50 SNU. With gallium they could have access to the whole spectrum of neutrinos, and what's more, with a good intensity. If a gallium experiment could be mounted, the answers would surely be found.

During the next five years the proposal was reviewed by various committees acting on behalf of the US Department of Energy, who funded Brookhaven Laboratory and much of the research in physics. The reports were uniformly favourable; the politics were not.

Physicists said that the experiment had immense potential for understanding the fundamental nature of the Sun, and should indeed be funded; the sting in the tail was that the money should come from the astronomy budget rather than from physics. Astronomers also regarded it as superb physics, but with the recommendation that the physicists should pay for it! The nuclear and particle physics sections of the Department of Energy could not get their respective constituencies to agree on the financial responsibility.

With the Department of Energy unable to get anyone to sign the cheque, Davis and Bahcall turned to the National Science Foundation. Here there was an immediate catch-22: Davis worked at the Brookhaven National Laboratory, and the NSF as a matter of policy does not support research proposals coming from laboratories, such as Brookhaven, which are already funded directly by the DOE. In desperation, Bahcall, based at the Institute for Advanced Study in Princeton, was elevated to principal investigator with his university address opening the doors to a request to the NSF. Here too the application ran into conflict over who ought to be underwriting it.

The end of a long story is that no major gallium experiment was ever funded in the USA. They did manage a pilot trial with about a tonne of gallium, which demonstrated the feasibility of the technique. Some of the insights and equipment developed in this preliminary study were eventually used in the large scale 'GALlium EXperiment', known as GALLEX, which involved European scientists and was mounted under the Gran Sasso mountain in central Italy.

Attitudes were very different in the Soviet Union. Moissey Markov, head of nuclear physics at the Russian Academy of Sciences, was so enthused that he helped to realise Pontecorvo's

1974 plan by setting up the Baksan neutrino observatory under the Caucasus mountains in Russia. Most important, Markov successfully negotiated the use of 60 tonnes of gallium, free of charge, for Russian physicists to use for the duration of their experiment. This led to a Russian–US collaboration known as SAGE, which stood for Soviet American Gallium Experiment. However, by the time it was begun, the Soviet Union was no more and the experiment's name was altered to Russian American Gallium Experiment, though prudently the acronym SAGE remained unchanged.

Those 60 tonnes that went to SAGE represented the total world supply of the element at that time. It would take two further years to produce a further 30 tonnes, which went into GALLEX. So by the 1990s the stage was set for two experiments, using gallium detectors, to look for the first sighting of the fundamental neutrinos from the Sun's primary fuel, the *pp* chain.

But even as preparations began, a new way of detecting solar neutrinos was being born, one that would revolutionise the field and create a new branch of science: neutrino astronomy.

Neutrinography of the Sun

By the beginning of the 1980s, Davis's experiment using chlorine in the Homestake mine was still the only one looking for solar neutrinos. Because of the difficulty of the experiment, and its reliance on the radiochemical technique, many physicists remained uncomfortable. True, for 20 years no one had found an error in either the experiment or the calculations – Bahcall was always very positive about that – but the calculations were complicated, involving lengthy computer codes, and while Bahcall always fully answered any questions that others put to him, there remained a nagging worry that something in the programs could be wrong. Bahcall was regarded by many as 'the guy who wrongly calculated the flux of neutrinos from the Sun'.[xxiv]

What was needed was a way of capturing neutrinos one at a time as they happened, rather than accumulating and inferring on a monthly basis after the event, as had been the case for Davis's radiochemical detector. This is what scientists working in the Kamioka mine in Japan realised they could do, and what other scientists working on the IMB experiment[34] below Lake Erie discovered by accident that they could do also.

Both of these experiments had been looking for signs that protons decay, because some theories attempting to unite the forces of nature implied that is what should happen occasionally. The stability of matter shows that if this happens at all it is exceedingly rare, the half life of a proton being many, many times greater than the life of the observable universe. To have any chance, they had built huge tanks of ultra pure water surrounded with thousands of photo-multiplier tubes or PMTs to catch any particles produced when protons decayed. PMTs act like light bulbs in reverse. When electric current enters the bulb of a lamp, it gives off light; when light enters a PMT, its energy is converted into an electric current, which can be sent to a computer that records the event. Where would the light come from in these tanks of water deep underground? The answer, so they hoped, was particles flying through the water at superluminal speeds.

---

[34] IMB stood for the universities of Irvine (California), Michigan, plus the Brookhaven National Laboratory, the three collaborating institutions. Fred Reines was the lead member from Irvine.

This needs a moment of explanation as surely nothing can travel faster than light? That is true in a vacuum, but light is slowed down when it passes through materials such as glass or water. It is then possible that charged particles, such as electrons, can travel through the water faster than the light can (though still of course slower than nature's ultimate speed limit, set by light in a vacuum). When this happens, there is a luminous analogue of a sonic boom, and a cone of pale blue light radiates out centred round the flight path. This is known as Cerenkov radiation after the Russian scientist, Pavel Cerenkov, whose experiments led to the understanding of the phenomenon.[35] When this cone hits the walls of the water tank, it is detected by the PMTs. The computer then reconstructs the shape and size of the ring, from which it is possible to infer both the energy of the original particle and the direction that it was moving.

The interest in proton decay had been driven by an erroneous experiment that had made people think that in a kiloton of material one proton could decay each day. Had this been true it would be possible to detect it so long as the background from cosmic rays could be eliminated. This led to a rush to build experiments in deep mines and tunnels underneath mountains.

Even so there was the problem of cosmic rays hitting atoms in the upper atmosphere and spawning neutrinos, which would penetrate the rocks and trigger the detectors at a similar rate to the hoped-for signals from decaying protons. Abdus Salam, a Nobel Laureate and enthusiastic theorist who believed strongly that protons decay, wrote a paper in which he suggested that

---

[35] Cerenkov radiation was first seen in the early 1900s by the Curies, and first studied by the French scientist L Mallet in 1926. However, he could not identify its nature, and so history, and the Nobel Prize committees, have honoured Cerenkov.

this unwanted background of atmospheric neutrinos could be eliminated if an experiment was done on the Moon. Don Perkins, a leading neutrino experimentalist and no respecter of theorists, reviewed Salam's paper with the comment that if Salam or other theorists wanted to go to the Moon, why not – 'the more the merrier'.[xxv]

It was only later that it finally dawned that if protons decay at all, the phenomenon is so rare as to be invisible. With this unfortunate realisation making these huge detectors deep underground useless for their original purpose, the teams of experimentalists started to study the neutrinos that hitherto had been the unwanted background. This was not an inspired choice: there was nothing else that could be done with these expensive behemoths.

The water-detector technique could be used to detect neutrinos, but required some modifications. The original experiments had been designed to find proton decay, but any particles coming from a decaying proton would have had much higher energy than a solar neutrino. Having found no sign of decaying protons, and aware of the growing tension surrounding the solar neutrino problem, the Kamioka team retooled the detectors so as to be sensitive to these lower energy neutrinos.

By chance, scientists were stumbling towards the place where real discoveries awaited.

Neutrinos from the boron-8 reaction in the Sun, the ones that Davis was studying, have enough energy that when they hit an electron in the water, the charged particle will recoil in the same direction that the neutrino had been travelling, like the head-on strike of a billiard ball.

The cone of Cerenkov radiation, which results as the electron rushes through the water at a super-luminal velocity, is then

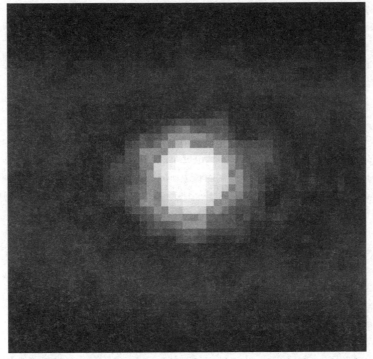

Figure 9 Neutrinography of the Sun. The Sun as 'seen' when its neutrinos are detected.

detected by the PMTs surrounding the tank. The size of the ring varies with the speed of the electron, which in turn depends on the energy that the original neutrino gave it. So these 'water Cerenkov' detectors had the possibility not just to detect neutrinos, but also to measure their energy and the direction that they had come from. Finally, the time that this happened is recorded.

The power of this detector was that all the information about the neutrino is known: its energy, when it hit, and where it came from. In particular you could confirm if they were coming from the direction of the Sun and not the result of some other source such as radioactivity in the surroundings. In effect what they had built was a neutrino telescope, a new window on the universe.

As a result they were able to make a 'neutrinograph' of the Sun. In this image the Sun appears to be much bigger than what we see with our eyes. This is because the directional accuracy of neutrino observation is not as good as can be done optically. Neutrino astrophysics is still in its infancy; images of the Sun and other astronomical objects in the future will become much sharper.

The team in the Kamioka mine completed the revisions of the Kamiokande[36] detector at the end of 1986. Then they had a stroke of luck. On 23 February 1987, utterly without warning, a supernova was seen to have erupted in the Large Magellanic Cloud, a satellite galaxy of the Milky Way in the southern skies. A blast of neutrinos from this explosion had been travelling across space for 170,000 years and passed through the Earth during about 15 seconds that day. The story of this will be told in chapter 10, but for now it is sufficient to say that both Kamiokande and the IMB experiments detected a handful of neutrinos from the supernova. This was the first and, so far, the only time that neutrinos from such an event have been detected. It is ironic that neutrinos from a supernova, and one from outside our galaxy no less, were seen and recognised even while the solar neutrino mystery was still being debated. After this singular event, IMB continued to look for evidence of decaying protons, while Kamiokande set about focusing on the new field of neutrino astronomy.

[36] Kamiokande stood for Kamioka Nucleon Decay Experiment.

Neutrinos Missing Everywhere

From 1987 to 1995, Kamiokande detected solar neutrinos. The Cerenkov light was only cleanly measured if the electrons had energies at least five times what Davis's detector could record. This meant that they were detecting only the relatively high energy solar neutrinos coming from the boron-8 reaction, but with much more detail than Davis had been able to achieve. In particular, they could measure the amount of energy that each neutrino had. The results showed that the number of neutrinos arriving from the Sun died away as their energy increased, as it would if these were neutrinos from boron-8, which is what Bahcall had predicted should happen. At last something from the solar model agreed with what was being seen. However, the total number stubbornly refused to fit, still amounting to only about a half of what was being predicted.

First Davis, and now Kamiokande: the rare neutrinos produced at the end of the *pp* cycle were too few in number. That much was now certain. The question of whether this would be true for the dominant neutrinos, the lowest energy ones produced in the earliest stage of the solar fusion cycle, remained unanswered by all this. That was what the SAGE and GALLEX experiments using gallium detectors, were ready to investigate.

SAGE began first, having monopolised the world's gallium, and GALLEX got under way in 1991. By 2000, SAGE had made nearly one hundred measurements of the solar neutrino flux over a period of ten years. The solar models predicted that they should see a huge number of neutrinos, around 130 SNU. When all the data were assembled, both SAGE and GALLEX found the same answer: the number was only about 70 or 80 SNU. So here again

117

the shortfall in solar neutrinos was about 50%. The results vindicated Davis: at all energies it seemed that the number of neutrinos being detected was only about one-half of that predicted by the solar models.

Whatever the reason was, it was not due to any failure by Davis's team; these independent experiments were all giving a common message. Nor did it look as if the Sun was the culprit either. By this time astronomers had made many measurements of the surface of the Sun, which showed how the Sun vibrates, and in turn gave sensitive information about its interior. These data on 'helioseismology' – the solar analogue of earthquakes – increasingly confirmed that Bahcall's assumptions about the inner workings of the Sun were correct. The evidence that the neutrino was the prime suspect was mounting.

In 1996, after a year of reconstruction, Kamiokande was ready for more work. With ten times more water and PMTs than before, the detector was renamed SuperKamiokande, or SuperK for short. If there was still any doubt that neutrinos were the key to the puzzles, SuperK was about to remove it. Neutrino astronomy could detect neutrinos not only from the Sun, and from a supernova, but also from the atmosphere. Cosmic rays hitting the upper atmosphere produce showers of neutrinos. SuperK started to detect these. While the measurements of solar neutrinos had turned out to be a surprise, that was nothing compared with what SuperK was about to find when it started observing the atmospheric neutrinos.

9

'I FEEL LIKE DANCING, I'M SO HAPPY'

Atmospheric Neutrinos

Neutrinos are produced in many circumstances, most around here having been born in the Sun, or in the rocks beneath our feet. In addition, large numbers of them come from cosmic rays.

Far above our heads, particles smaller than atoms are showering down from the heavens. They are the result of stars that exploded long ago. Electric and magnetic fields permeate interstellar space, and whip the debris into violent motion. Some of these particles have energies far higher than anything that we can reproduce in experiments on Earth. As we travel through the cosmos, these cosmic rays are continuously hitting us head on.

The violence of these collisions breaks atoms in the air into little pieces, creating showers of secondary particles moving almost at the speed of light and in the same direction as their

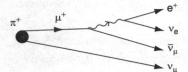

Figure 10 One electron-neutrino for two muon-neutrinos. *See also fig 6* A positively charged pion (π^+) decays into a positively charged muon (μ^+) and a single neutrino. This is a muon-neutrino denoted v_μ. The muon in turn decays into a positron, a muon-antineutrino denoted $\bar{v}_\mu$ and an electron neutrino denoted v_e. The end products of the pion decay consequently contain two neutrinos of the muon type for every one of the electron-type.

original parents. The vast majority of the cosmic rays are absorbed by the air or the ground, and never penetrate to the deep underground caverns where the neutrino detectors await. However, the rays contain pions and muons, many of which decay before they are absorbed. As a result they produce neutrinos, which reach the detectors because the intervening rocks are effectively as transparent to these as to solar neutrinos.

There are two major differences between the neutrinos that are the progeny of cosmic rays hitting the atmosphere and those that have come from the Sun. First the 'atmospheric' neutrinos have energies tens to many hundreds of times greater than their solar cousins. Second, whereas the Sun produces 'electron-neutrinos', v_e, the debris of cosmic rays produces predominantly the 'muon-neutrino' variety, v_μ. Experiments over many years in laboratories such as CERN had shown how the cosmic particles behave, and from these it was clear that the neutrino content of the debris would be one electron-neutrino for every two of the muon-variety, the latter being one v_μ and one antineutrino, denoted $\overline{v_\mu}$.[37]

A crucial feature of the water detectors is that in addition to recording the energy of a neutrino and where it comes from, they can also tell what variety of neutrino they have captured. When a neutrino hits an atom it picks up an electric charge, becoming either a muon or an electron of either positive or negative charge. As we have seen on page 113, this charged particle flies through the water, radiating Cerenkov light as it goes, which is then detected by the PMTs. Muons are some 200 times heavier than electrons and punch through easily, travelling in straight lines, whereas lightweight electrons are knocked from their path. While this is happening, they radiate energy, which spawns further electrons and positrons. So a single electron generates a shower of charged particles, which is spread diffusely around the original direction of its forebear, the neutrino. The resulting patterns of Cerenkov light for electrons and muons are different. For the muons there is a sharp ring whereas for the electrons, which have been scattered this way and that, the ring is fuzzier. The shape of a ring, and the precise times of arrival of the Cerenkov light at the PMTs around the ring's circumference, reveal the direction and energy of the incident neutrino as well as its variety.

In the mid 1980s there had been hints in both IMB and Kamiokande that the ratio of muon-neutrinos to electron-neutrinos coming from collisions in the atmosphere, was nearer to one than the expected value of two. This became known as the

---

[37] The initial debris contains many pions, π, with slightly more carrying positive charge, π^+, the charge being the 'memory' of the positively charged nuclei that helped spawn them. When π^+ decays, it almost always produces a positively charged muon and a muon-neutrino. The muon in turn decays leaving a positron, a v_e and a $\overline{v_\mu}$.

atmospheric neutrino anomaly. Speculation began to grow that the neutrinos, rather than the solar astrophysics, might be the cause of the anomaly in Davis's experiment. No one could be sure whether the anomaly with atmospheric neutrinos indicated too many ν_e or too few ν_μ, or whether it was an artefact of the detection method. This is what had driven the upgrade at Kamioka that converted Kamiokande into SuperK.

SuperK was larger than its predecessor, with the PMTs on its surface covering an entire acre. Its electronics were specially tuned to get the maximum information about the neutrinos. It was possible to tell whether they had come from the sky, 20 km above Japan, or instead from the far side of Earth, travelling 13,000 km through the planet and then up through the bottom of the tank. They could measure the directions well enough to tell from where around the globe they had originated.

The new detector captured many more muon-neutrinos than before. The increased number of data showed that the number of muon-neutrinos and electron-neutrinos ended up comparable because muon-neutrinos were disappearing. This was interesting, but the data revealed an even more remarkable fact: the deficit was greater for the neutrinos coming through Earth than for those arriving from overhead. The more data that SuperK accumulated, the clearer the message became. Everything was consistent with the idea that the cosmic rays indeed produce the muon and electron varieties in the ratio of two to one but that the further the muon variety had travelled, the more likely it was that its members had disappeared.

Immediately everyone thought about Davis's problem with the electron-neutrinos from the Sun. Could neutrinos disappear in flight, not just muon-neutrinos as in the atmospheric anomaly, but electron-neutrinos too? After all, they had travelled 150 million kilometres from the Sun, and if it was possible for atmospheric neutrinos to do a disappearing act, there would be plenty of time for the solar ones to do likewise. For the first time, even the sceptics began to agree that, after all, Davis's results might be correct. The Sun is behaving fine; it is the neutrinos that are behaving oddly. The question now was: if neutrinos disappear, what becomes of them?

Oscillating Neutrinos

Half a century after Pauli had first proposed his 'desperate remedy', experiments had proved not only that the neutrino exists, but that it occurs in three distinct varieties. This intense and continuous pursuit of neutrinos had gone in parallel with the solar neutrino problem, which, far from refusing to die, continued to have new life. Davis's experiment, as we have seen, became more accurate, and the data ever more insistent that there is a shortfall of solar neutrinos, at least as detected by his set-up.

Given the amount of time and resources invested, and the number of scientists who by the 1990s had sweated over this conundrum, it is ironic that two decades earlier, within a year of Davis's first tentative announcement, the answer had been found – and ignored. Following hot on the experimental demonstration in 1962 that the electron-neutrino and muon-neutrino differ, Bruno Pontecorvo and his colleague Vladimir Gribov, in Russia, and Maki, Nakagawa and Sakata in Japan[xxvi] had realised that the existence of more than one type of neutrino offered a possible solution to the solar neutrino problem.

The Sun produces neutrinos of the electron type, v_e; the chlorine detector records the arrival of neutrinos only of this same type. If nothing has happened along the way to change the v_e into something else, the accounting at Davis's detector here on Earth tells you how many v_e were created in the Sun less than ten minutes ago. However, by 1962 it had been established that there is another variety of neutrino, the one with muon affinity, v_μ. This led to an intriguing question.

Schwartz, Steinberger and Lederman had shown that neutrinos remembered their provenance over tens of metres, but what if this memory failed over larger timescales? A few nanoseconds traversing a laboratory is a mere trifle; neutrinos coming from the Sun have been travelling for nearly ten minutes, over a billion times longer. Could a v_e that was born in the Sun manage somehow to change its spots, turning into a v_μ during its journey through space?

If the electron-neutrino switched its identity like this, it would pass clean through Davis's detector as if nothing were there. In effect it would have disappeared without trace.[38] Only those electron-neutrinos that survived the journey unscathed would be caught in Davis's trap and counted. This could explain the 2–3 SNU recorded by Davis versus the 6 that Bahcall had calculated should be there.

The idea that neutrinos have some sort of personality disorder, being produced in the Sun in one state and managing to change identity in their travels, ran counter to everything in the textbooks. According to the standard theory of particle physics, this was impossible. Or at least, it was impossible if, as everyone believed, neutrinos are massless and travel through space at the

---

[38] Muon-neutrinos can only turn into muons if the neutrino has enough energy. This can happen for neutrinos produced by the high energy cosmic rays, but those from the Sun have too little energy. So solar neutrinos that have metamorphosed into the muon variety are in effect invisible.

speed of light. Long before Davis had discovered the shortfall in solar neutrinos, Pontecorvo had noticed that the laws of quantum mechanics allowed neutrinos to oscillate back and forth between one state and another, but only if they had some mass. It didn't need to be large; in fact, it could be, and probably is, triflingly small, thousands of times smaller even than the mass of an electron. After Davis's announcement in 1968, the following year Gribov and Pontecorvo published their theory, based on the hypothesis that there are two varieties of neutrino, with different masses.[39]

In quantum mechanics, certainty is replaced by probability, which rises and falls like a wave. The wavelength depends on the speed and mass of the particle. So the waves for two particles having the same energy, but slightly different masses, will have marginally different wavelengths. The weirdness of the quantum world goes even deeper because it allows the electron-neutrino, which the Sun produces, to be a hybrid of neutrinos with two different masses. As this rushes across space, the quantum waves associated with these two states wobble at different rates.

In effect there are two waves, of different wavelengths, which are travelling along and interfering with one another as they go. As two sound waves of slightly different frequencies mingle and give a pulsing beat of intensity along with the average note, so the quantum waves associated with two neutrinos of slightly differing masses can give an analogous rise and fall in intensity. The result is that only occasionally along the journey do the two waves match up in the precise form that they started out. It is only at these points that they reconstruct to represent an electron-neutrino. Everywhere else the shape of the wave subtly

---

[39] Only two varieties were known at that time. Their idea generalises immediately to the case of three.

oscillates, such that, in effect, a mixture of electron-neutrino and muon-neutrino is present. When something such as an atom of chlorine in a tank 1400 metres beneath the hills of Dakota gets in the way and interrupts the flow, the quantum wave miraculously converts into a v_e or v_μ. It is impossible to know which it will be. All that quantum theory implies is a probability of finding the one or the other. In effect, if you do this enough times, on average it will be roughly 50:50 an electron-neutrino or a muon-neutrino.

The phenomenon is reminiscent of Escher's drawings of 'Metamorphosis' where, as you traverse the diagram, you perceive one animal to be gradually changing into another. As an example, imagine some weird hybrid that can metamorphose between a cat and a dog. The dog sets out from its home and walks along the street, transforming into a cat as it goes. Halfway along the block the transformation is complete. The former dog (now a cat) continues walking and metamorphosing. By the end of the block it has returned to a dog once more. When you look at the dog-cat, what you see will depend on how far along the block you are.

Now suppose that you are not receptive to dog-cats, only to things that are one or the other: a dog or a cat. If you are near the start or the end of the block, you will most likely interpret it as a dog. If you are near the midpoint, you are more likely to interpret it as a cat. If your eyes are only capable of seeing dogs, and not cats, then you might conclude that the doginess has disappeared relative to what had started out.

So it is with the neutrinos. In this analogy, the electron-neutrino is the dog and the muon-neutrino the cat. The Sun emitted a dog, and Davis's detector was a dog-catcher. In this theory, there was nothing wrong with the Sun; it was the neutrinos that were the culprits. Davis's tentative measurement of an apparent shortfall of solar neutrinos could be understood. All that was required was to give up the standard model of particle physics, which included the assumption that neutrinos are massless and travel at the speed of light. Not surprisingly, few were prepared to do so, and the idea was widely regarded as little more than a mathematical curiosity.

However, not everyone had ignored the idea of neutrino oscillations. For oscillations to happen, neutrinos could not all be massless. This did not violate any sacred principle, and Murray Gell-Mann, one of the most influential theoretical physicists of the 20th century, even opined that what Nature does not forbid, will happen. With the arrival of the atmospheric neutrino anomaly, and Davis's ever-improving data about solar neutrinos, people began to wonder if neutrinos might indeed have mass, even though trifling compared to those of all other material particles.

In Pontecorvo's original theory, electron-neutrinos that had been created in the centre of the Sun could convert into muon-neutrinos, or even tau-neutrinos, which were invisible in Davis's experiment. However, if the solar models were correct, and Davis's experiment was also, then a large fraction of electron-neutrinos would have to have oscillated away into one of the other forms. In order to do so, neutrinos would have to be so schizophrenic that naming them electron- and muon-neutrinos would be perverse. This was one of the reasons that Pontecorvo's idea had been widely ignored.

Opinions started to change when three theorists discovered a novel implication of the oscillation idea. This became known as the MSW effect, the acronym being their initials. The American, Lincoln Wolfenstein, in 1978, and two Russians, Stanislav Mikheyev and Alexei Smirnov, in 1985, had realised that as neutrinos passed through the layers of the Sun, the presence of matter could amplify the likelihood that neutrinos oscillate, provided that their masses were in a particular range. As a result of this increased chance, the neutrinos needed only to be mildly confused about their identities at the outset. During their journey to the surface of the Sun, even a small amount of personality confusion could grow into a full blown identity crisis thanks to the presence of matter.

Largely because of the elegance of the theory, around 1990, physicists began to take the idea of neutrino oscillations seriously. The discovery of the atmospheric neutrino anomaly had begun to emerge around 1985, but it was not established until 1993 when the first results from SuperK showed that the further that muon-neutrinos travelled, so the more likely they were to disappear.

By 1998, SuperK was able to announce that the amount of the shortfall varied not just with distance but also with the energy of the atmospheric neutrinos. If their identities really oscillated back and forth, then according to the theory of relativity the oscillations should be faster for lower energy neutrinos than for those with higher energies, and that is exactly what the data showed. Everything was in accord with the hypothesis that neutrinos oscillate.

It had taken 30 years, but Pontecorvo had been vindicated. Would this also resolve the solar neutrino problem? Could Davis and Bahcall both be right after all?

SNO

SNO, the Sudbury Neutrino Observatory in Ontario, Canada, was designed to solve the solar neutrino problem once and for all. It was looking at one particular high energy branch of neutrinos, as had Davis. These were only about one hundredth of one per cent of the whole, but their higher energy gave them a higher chance of reacting. However, SNO would create a novel possibility: measuring not just neutrinos of the electron type, but of all types. This would involve initially a comparison between SNO's results and those of SuperK. If the Sun's neutrinos had indeed turned into muon or tau types by the time they arrived at Earth, SNO would be able to prove it.

In order for the experiment to work properly, the SNO detector had to be as big as a ten-storey building. Its unique feature was its one thousand tons of heavy water, in which the hydrogen is replaced by deuterium, a nucleus consisting of a single proton and a neutron. This was loaned to the collaboration by Atomic Energy of Canada Ltd, and filled a 12-metre diameter acrylic plastic container. In turn, this was surrounded by a geodesic sphere extending to a diameter of 18 metres, with ten thousand light sensors around the surface. The whole apparatus was located within a cavity, filled with ultra-pure ordinary water, 34 metres high and 22 metres across, and this was placed two kilometres below the ground in an old nickel mine beneath Sudbury, Ontario. Even a single teaspoonful of dust in the whole apparatus would have rendered it useless.

When measurements began in 1999, SNO had taken nearly ten years to construct at a cost of 73 million Canadian dollars. This included the fee of one dollar, which was paid to the Canadian Atomic Energy company for the loan of the heavy water – valued at 300 million dollars! Neutrino astronomy was becoming 'big-science'. The SNO detector was able to sense electron-neutrinos in a similar way to what SuperK had done. The collisions of electron-neutrinos produce electrons, which give off Cerenkov radiation, blue light, as they travel through the water.

As in SuperK, the intensity of the light depends on the energy of the electron, from which the range of energies of the incoming neutrinos could be determined. SNO intercepted about ten neutrinos a day.

The first results were announced on 18 June 2001. They were trailed as providing the 'solution to a 30-year-old mystery – the puzzle of the missing solar neutrinos'.

The flux of neutrinos of the electron type turned out to be 1.75 million per square centimetre each second. Then Art McDonald, the leader of the SNO team, announced that they had done something novel. The SuperK detector in Japan had also measured electron-neutrinos in a similar setup but in addition had some sensitivity to neutrinos of the muon or tau type. Combining the results from SNO and SuperK made it possible to estimate not only how many electron-neutrinos had reached Earth, but also how many of all varieties of neutrinos were arriving. In the same units as SNO's 1.75 million electron-neutrinos, SuperK had measured 2.32 millions in total. The difference was because SuperK sampled some muon or tau-neutrinos as well.

Already this was enough to show that some of the electron-neutrinos must be changing into other varieties during their journey from the Sun. The challenge now was to determine how many muon or tau-neutrinos SuperK was intercepting in total. When the calculations were all done, the combined result from SNO and SuperK showed that the total neutrino flux, of all varieties, added up to 5.44 million per square centimetre each second, with an uncertainty of about twenty per cent lower or higher. This number agreed with what Bahcall had predicted.

So, by 2001, the shortfall in solar neutrinos first seen by Davis had been confirmed by four other experiments. SAGE and

GALLEX were finding reduction in the lower energy neutrinos, and the combination of SuperK with the first results from SNO showed that the shortfall was because the neutrinos of the electron type were only about one-third of the whole. The implication was that electron-neutrinos change to the other varieties of neutrino, which in turn can themselves transform from one variety into another. Over the 150 million kilometres from the Sun to Earth, the distribution of neutrinos settles down to a more or less even mixture of all three varieties. The uncounted neutrinos weren't missing at all: en route they had changed into forms that were simply much harder to detect.

This was a major result, but there remained a weakness: the conclusion relied on combining the separate data from two quite different experiments. The ideal would be if a single experiment could provide the lot. That is what the SNO team set out to do next. During 2002/3 they gathered more data, strengthening their results and making them more precise. A clever development was using the heavy water in a way that enabled all varieties of neutrino to be revealed in the SNO detector, whereby there would be no need for comparisons with data from other experiments. As one member of SNO wryly said, 'This one will be watertight'.

This was the idea originally proposed by Herb Chen. In heavy water, the hydrogen atoms are replaced with heavy hydrogen – deuterium – each of whose atomic nuclei consists of a proton bound to a neutron. The neutrinos had enough energy to split the deuterium nucleus in two, releasing its individual neutron and proton. What emerged would depend on which type of neutrino had struck. An electron-neutrino could pick up charge, turning into an electron and converting the neutron into a proton: an electron and two protons would result from that collision. Muon or tau neutrinos could not do this. However, they – and also the electron-neutrino – could bounce off the proton or neutron, kicking it out from the deuterium but leaving the proton and neutron otherwise unchanged. By comparing the number of examples of the latter category with the tally where two protons emerged, SNO could measure the flux of all varieties of neutrino, and in addition determine what fraction were electron-neutrinos. Sadly, Herb Chen died of leukaemia in 1987, and so didn't live to see the result of his inspired idea.

A major improvement occurred in the summer of 2002, when they added 2 tonnes of high-purity table salt, sodium chloride, to the 1000 tonnes of heavy water at the heart of the detector. The chlorine in the salt increased the chance of intercepting neutrinos and further helped to discriminate between the different types.

In an interim report, in 2002, they announced that the results from SNO alone were already good enough for them to say that they were 99.999% confident that neutrinos from the Sun change from one type to another before reaching the Earth. Finally, on 7 September 2003, they announced the definitive results. The number of neutrinos of the electron type was the same as they had found before: 1.75 million per square centimetre each second. As for the total number, that came out to be 5.21 million. This agreed with the earlier result, which had combined SNO and SuperK measurements. It was also more precise: electron-neutrinos are close to one-third of the whole.

The conclusions at last were clear. First, that for 30 years Davis had been correctly measuring the number of solar neutrinos that are still electron-neutrinos when they reach Earth. Second, and immensely exciting for the astrophysicists, was the confirmation that Bahcall's calculation of solar neutrino production was correct. Bahcall commented that the agreement was 'so close that it was embarrassingly close.'

For three decades, people had doubted him. Bahcall, who, as we mentioned earlier, said that he was regarded as 'the guy who wrongly calculated the flux of neutrinos from the Sun', had suddenly been proved to have been right all along. He later compared it to being a person who has been wrongly accused of some heinous crime, until a DNA test proves that he is not guilty: 'That's exactly how I felt.' On receiving the news, he had responded more spontaneously, 'I feel like dancing, I'm so happy'.

Bahcall's successful calculations showed that the numbers of neutrinos are sensitive to the temperature at the heart of the Sun, multiplied by itself 25 times. The result of all this was that measurement of the neutrino flux is a sensitive thermometer for the Sun's nuclear fusion furnace. It is truly remarkable that detecting a flash of light at the bottom of a deep mine can measure the temperature at the heart of the Sun.

Bahcall and Davis had been vindicated. This was the moment when a new science, 'neutrino astronomy', started to become a real possibility. Neutrino astronomy had begun with the search for solar neutrinos and had become a quantitative science with the SuperK and SNO experiments. By the time that the solar neutrino mystery had been solved, neutrinos from beyond the galaxy had also had ten seconds of fame with a piece of pure luck. It would be this that brought home the potential power for discovery that neutrino astronomy might offer.

10

EXTRAGALACTIC NEUTRINOS

Supernova

Where were you at 07.30 GMT on 23 February 1987? I was
having breakfast when, unknown to me, a burst of neutrinos
passed through my cornflakes. All the time, we are being bathed
in the flux of solar neutrinos, but the sudden burst that February
morning was quite different. It was a blast from a dying star,
170,000 light years away in the Large Magellanic Cloud, or
'LMC', a satellite galaxy of our own that is visible in the southern
skies. For over 25 years, astrophysicists had believed that the
gravitational collapse of a supernova is a copious source of neutrinos. In fact, they argued that the brilliant flash of light, the traditional manifestation of a supernova that can briefly outshine an
entire galaxy, is only a minor part of the drama. Powerful though
this intense electromagnetic radiation is, the visible light, radio
waves, X-rays and gamma rays all add up to less than one per cent

of the whole. The bulk of the energy radiated by the supernova is carried away by neutrinos.

These neutrinos were invisible in the past, but not now that we have neutrino telescopes. The exciting news was that, in this case, for the first time, we detected neutrinos emanating from outside our galaxy, and proved that the theory of a supernova is right: when stars collapse they throw off their energy as neutrinos, up to 10^{59}, a hundred billion trillion trillion trillion trillion of them in all.

The fact that neutrinos were detected from this supernova at all is a fortunate coincidence, and could not have been planned for. The last supernova visible to the naked eye was in 1604, since when, none had been seen for more than three centuries until this one burst into prominence in 1987.

Actually, the violence really took place in the Large Magellanic Cloud 170,000 years ago. A flash of light and a blast-wave of neutrinos flew out from the debris. Travelling 10 million miles each minute, they raced away from the site, left the LMC and headed out across intergalactic space, their 1987 rendezvous still far in the future.

Ahead of them lay the Milky Way, in an arm of which, on the small planet Earth, human life had advanced to the stone age. The shell of radiation travelled onwards for over 165,000 years. By this stage, 3000 light years away, people around the Mediterranean were beginning to be aware of the heavens and were inventing science. By the 1930s, their descendents were beginning to suspect that radioactive processes spawn neutrinos, though it was doubted whether anyone would ever detect one.

Meanwhile, the wave from the collapsed star was approaching the Earth through the southern heavens. It was 31 light years away from Earth when Clyde Cowan and Fred Reines cleverly proved that neutrinos exist. The blast wave was still 23 light years away when Ray Davis started operating his solar neutrino detector in the Homestake mine. Although able to detect neutrinos coming from the Sun, it would have been almost blind to any from a supernova. However, while the approaching neutrinos were only a light year away – just one part in 170,000 of their journey – scientists in America and Japan had just finished building huge tanks of pure water underground, designed to look for signs that protons decay.

The Japanese detector, Kamiokande, the forerunner of SuperK, contained 3000 tonnes of pure water. A similar story was taking place 600 metres beneath the bed of Lake Erie in Ohio, where a team from the Irvine campus of the University of California, the University of Michigan, and the Brookhaven National Laboratory, known as IMB (Irvine, Michigan, Brookhaven) was also looking for proton decay with a tank containing 7000 tonnes of water.

While neither of these experiments found any examples of decaying protons, they each turned out to be sensitive to neutrinos whose energies were higher than those coming from the Sun. What no one realised was that a shell of these particles was heading their way at the speed of light.[40] The 10^{59} neutrinos that had set out 170,000 years earlier had, by this time, spread out over the surface of a sphere whose radius was 170,000 light years, big enough to encompass the entire galaxy. The thickness of the shell was about ten times the distance from the Earth to the Moon, the consequence of the few seconds that it had taken for the neutrinos to diffuse out from the collapsed star, whose density

---

[40] As neutrinos are now known to have a small mass, they actually travelled marginally below the speed of light, but the difference is negligible.

was like that of a huge atomic nucleus. If you work out the density of 10^{59} neutrinos spread over a shell that is 340,000 light years across, you find that it is similar to what we receive continuously from the Sun. Considering how intense solar neutrinos are, and that they are born only 8 light minutes away rather than 170,000 light years, one begins to get a sense of how remarkable a supernova stellar explosion really must be. Even more impressive is that each neutrino from the supernova can have ten to a hundred times more energy than the dominant ones from the Sun.

At last, on 23 February 1987, they hit the Earth, passed right through, and carried onwards into northern skies. In a matter of seconds, a thousand trillion neutrinos passed through IMB, a similar number through Kamiokande and, we can be sure, through the Homestake mine and other laboratories too. However, only IMB and Kamiokande were sensitive to them. Out of these hordes, only eight carried enough energy to be detected in IMB, and eleven were seen in Kamiokande.[41] These detectors in Japan and the USA were in the northern hemisphere whereas the Large Magellanic Cloud is only visible to the eye in the southern skies. The neutrinos from the supernova had passed right through the Earth and entered the detectors from below.

One of the great advantages of IMB and Kamiokande was that both the direction and the energies of the neutrinos could be measured. Whereas most of those coming from the Sun have less than 1 MeV energy apiece,[42] and while the highest energy ones that Davis's Homestake experiment with chlorine was sensitive

---

[41] Actually they were antineutrinos.

[42] To recap, 1 MeV = a million electron-volts, or about one ten-trillionth of a joule.

to had at most 14 MeV, those detected from the supernova were in the range 10 to 50 MeV. Kamiokande's eleven were in the 10 to 20 MeV range, while IMB – which was not sensitive to these relatively 'low' energies – recorded eight between 20 and 50 MeV apiece.

The amount of information that scientists were able to deduce from just these few events, was remarkable. First, the energy. The numbers actually captured were but a tiny fraction of the trillions that would have passed through the detector. In turn these detectors occupied a mere morsel of the surface of the vast sphere through which the blastwave was then passing. Take all of this into account and you can get an estimate of the total energy that neutrinos carried away from the supernova. It turned out to be about one-tenth of the total energy contained in the Sun, that is, one-tenth of its mc^2, and remarkably in accord with what the theories of supernova explosions had predicted. According to these theories, energy is also radiated both as light and in newly formed atomic elements; in addition, there is still a lot of energy trapped within the mc^2 of the compact neutron star that remains behind after the explosion. In total one begins to get a sense of how powerful a stellar explosion must be.

The discovery that the energies of the neutrinos from the supernova were much larger than those of the solar neutrinos immediately showed that the temperature in the star prior to its collapse was correspondingly hotter. It turned out to be about 40 billion degrees, which also fitted with what the theories had predicted. So the energy, the total number of neutrinos, and also the time duration of the blast, were all in accord with the supernova being the result of a star collapsing under its own weight.

The fact that the burst was spread over about ten seconds was very significant. Had the neutrinos come from the demise of a tenuous stellar object, everything would have been over in a thousandth of a second. However, to diffuse out from a very dense object, taking many seconds to escape from the surface, was what would happen if the object was as dense as an atomic nucleus. All this suggested that formation of a neutron star had occurred. This is what astrophysics theorists had suspected supernova explosions to be, but it was the first time that direct evidence had been found.

By detecting this momentary blast of neutrinos, humans had taken their first look into the workings of a supernova. In doing so, they confirmed everything that had previously been just theory: a supernova is the result of a star collapsing to form a neutron star. A melange of elements in the periodic table is formed in the process, including those needed to seed future life. This is as near as we have yet reached to confirming the belief that we are indeed made of stardust, or if you are less romantic, the products of an extinct nuclear reactor.

As these observations have confirmed the theorists' predictions that most of the energy produced in supernovas is radiated away in the form of an immense burst of neutrinos, the world's neutrino laboratories are eagerly awaiting the next one. The aim will be to measure not just numbers and energies, but also the variety of flavours that arrive here. When the star collapses, the density at its core reaches 10^{14} g/cm^3 – one hundred trillion grammes in every cubic centimetre. This is so high that protons and electrons combine to form a neutron, in the process releasing an electron-neutrino.

This is the first of two independent ways that neutrinos are made in a supernova. The neutron core that results has a

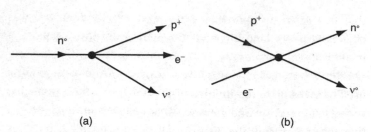

(a)   (b)

Figure 11 Neutron stars and neutrinos are made together. The story began with Fermi's model of neutrino decay (fig 2 repeated here as fig a). When a neutron star is made, this same basic process is at work. Electrons and protons in a dense star are squeezed so tightly together that they turn into a neutron and a neutrino (fig b). The neutrons form the neutron star; the neutrinos are radiated into space. The superscripts denote the amounts of electric charge.

temperature of over 100 billion degrees, and this thermal energy is dissipated via the second mechanism: the formation of neutrino–antineutrino pairs of all flavours. It is this second way that is predicted to be the most intense source of neutrinos from supernovae. A precise measurement of the ratios of flavours arriving from supernovae could enable the relative importance of the first stage (which produces electron-neutrinos) and the latter, which produces them all, to be determined.

Oscillating Neutrinos

'Neutrino astronomy' had begun when Davis first detected solar neutrinos, though it had taken more than 30 years before everyone was convinced. The arrival of SuperK, with its neutrinographs of the Sun, and its detection of neutrinos from the atmosphere and from a supernova, established neutrino astronomy as a new field of science. This offered the exciting prospect that by capturing neutrinos from more distant astronomical objects we may eventually be able to explore the universe at large in a novel and unique way.

From a pragmatic point of view, if neutrinos were ever to be used to investigate the innards of more exotic celestial objects, such as supernovae and gamma ray bursts, it would be imperative to understand the neutrinos themselves, for what has become abundantly clear is that neutrinos change their nature en route from the Sun. Electron-neutrinos certainly disappear, but whether they end up as the muon variety, or the tau variety, or a mix of both, has become the new question. Also, and most radical, was the implication that if neutrinos are indeed oscillating as Gribov and Pontecorvo had long suggested, this would mean that they cannot all be massless. Neutrinos with mass would lie outside the standard model of particles, which would need revision.

With the solar neutrino problem at last being resolved, new deep questions began to arise. How rapid are the oscillations; when one variety of neutrino disappears, which variety is created? To make neutrino astronomy a quantitative science, neutrino oscillations will first need to be understood.

When Cowan and Reines first detected the (anti)neutrino, they had to be near to the nuclear reactor that spawned them in order to capture a few. Today, with huge detectors underground, it is possible to pick up antineutrinos produced in reactors many kilometres away. This is fine if reactor neutrinos are what you want to study, but they can be an annoying background if your real interest is in neutrinos from elsewhere. An experiment in the 1990s, wanting to make a sensitive measurement of neutrinos, needed to know how many would also be coming from a nuclear power plant nearby. Finding out wasn't easy.

A call to the press officer drew a blank when he was asked how many antineutrinos the power station was producing. Possibly ignorant of the existence of the antineutrino at all, and sensitive to the ongoing protests by environmentalists who objected to a nuclear power station in the locality, the answer came abruptly: 'None!'

The scientist then explained that a nuclear reactor produces both power and antineutrinos inexorably together, therefore if there were no antineutrinos, then neither was it producing any power, so why was the company sending out electricity bills? Either the power company was defrauding the public or the answer 'none' could not be correct. The press officer promised to check and get back to the caller.

Later the phone rang in the scientist's office. It was the press officer. 'You're correct', he said, 'we do produce antineutrinos', and duly gave a number, adding fortissimo: '*but NONE escape!*' Unless the power plant had found the holy grail of how to capture each and every neutrino, this too was wrong, but at least the scientists had the number they needed. From this, a clever idea grew: why not use antineutrinos from reactors as the source of an experiment – to study antineutrinos! By detecting them far from their source, and comparing how many arrived with how many had set out, the idea was that it might be possible to tell if they had oscillated from one form to another en route.

That is what scientists in Japan have now done. They have used a detector in the Kamioka mine to measure the energies of antineutrinos. Known as KamLAND, for Kamioka Liquid Scintillator Neutrino Detector, it is sensitive to antineutrinos of only the electron variety, which is the form that the nuclear power plants produce. From the direction each antineutrino arrives, they can determine in which of 53 nuclear power plants in Japan it originated. By comparing the numbers arriving with what set out from the power stations, they found that the intensity didn't die away uniformly with the distance travelled. Instead, it depended on both distance and the energy, and in such a way that it fell away and then rose again. This turned out to be the key.

The greater the energy that an antineutrino has, the nearer to the speed of light it is travelling. When divided by the energy, the distance travelled is a measure of the time that the antineutrino has been in flight. The rise and fall turned out to depend on the flight-time. It had all the hallmarks of an oscillation.

KamLAND is over 100 kilometres away from some reactors, and it finds that on average about 40% of the anticipated number of antineutrinos have disappeared. There have been several experiments placed within about a kilometre of a reactor, but no signs of oscillation have shown up in these. In the Ardennes, and at Palo Verde in Arizona, the results agree: one kilometre is not enough for a neutrino to change its spots measurably.

KamLAND shows that neutrinos oscillate, but we still don't know enough to say which flavour of neutrino oscillates into what.

However, it is beginning to be possible to learn about neutrino masses. The mathematics of oscillations gives a measure of the differences in the mass-squared. This turns out to be a very small number, some 10^{-5} in units of eV^2. What does this mean for their actual masses? Well, that depends on how big one of them is. If one is indeed massless, then another must have a mass of about only 10^{-2} eV. For comparison, an electron weighs in at some half a million electron-volts.[43] If they are, however, both around 1 eV each, their masses can differ by only 10^{-5} eV.

---

[43] An electron-volt, or eV, is actually a unit of energy. The terms mass and energy are used interchangeably in physics. They are related by $E = mc^2$, where c is the speed of light.

In any event, one thing is certain: some varieties of neutrino must have mass. Possibly all of them do, but the values are exceedingly tiny, even on the scale of the lightweight electron. Why their masses are so similar, and yet not quite identical, is one of the major puzzles that future generations will hopefully answer.

MINOS

Among the lakes and forests of northern Minnesota, near the township of Soudan, there is an iron mine dating from the early 1900s, which followed an extremely pure seam of magnetite iron ore 800 metres deep into the Earth. By the 1950s it had become uneconomic to extract the ore, so the owners, US Steel, gave the mine to the state of Minnesota which now runs it as a state park with tourist trips to the underground workings. In the early 1980s, physicists from the University of Minnesota – who were looking for evidence of decaying protons – realised that this would be an ideal site for an underground detector. As a bonus, it could also record neutrinos that had been produced by collisions of cosmic rays in the atmosphere.

Together with colleagues from the USA and the UK, they built an experiment in an excavated cavern at the lowest level of the mine. The novelty was that it worked on a completely different principle from the water detectors. The charged particles produced by neutrino interactions were detected through the electrons that they liberated from the noble gas, argon. They found no evidence for unstable protons, but their detection of atmospheric neutrinos – those produced by cosmic rays hitting the upper atmosphere – confirmed the reports from the Kamiokande experiment that these exhibited a deficiency of muon-neutrinos. This showed the effect to be real, and not some artefact of the water detection technique.

In the late 1980s, Maury Goodman recognised that, at a distance of 735 km from the Fermilab accelerator complex outside Chicago, the Soudan mine offered the chance of systematic measurements of neutrino oscillations. The problem with atmospheric neutrinos is that physicists have no control over their production. You have to accept what chance provides, and you try to model how the neutrinos are produced from prior knowledge of the cosmic rays and the interaction processes. However, if the neutrinos are produced in an accelerator, they can be customised, enabling selection of either muon-neutrinos or electron-neutrinos, and with specific energies.

The accelerator first produces high energy protons, which are slammed into a carbon target. This produces large numbers of electrically charged pions, which are focused into a parallel beam and sent along an evacuated tunnel where they decay into muon-neutrinos. At the end of the tunnel, the rock-face filters out all of the charged particles leaving just a beam of neutrinos.

By pointing the tunnel at the Soudan mine, the beam would pass through the detector that was already there, and neutrino interactions could be studied to see if any neutrinos had indeed disappeared. Because of the curvature of the Earth, the beam had to be directed downwards at an angle of around 3°; the ensuing path of the neutrino beam is therefore completely through the Earth's crust.

From the Kamiokande and Soudan results it was expected that, if the deficit in neutrinos was due to oscillations, the muon-neutrinos would be oscillating to become tau-neutrinos. Since muon-neutrinos produce muons when they interact, and

145

tau-neutrinos produce tau particles, a detector that isolated muons could test whether muon-neutrinos had disappeared. Serendipitously, 735 km was just about right to match the peak of the expected oscillation for a Fermilab beam.

As always in neutrino experiments, the problem was that almost all of the neutrinos would pass through the Earth and the detector without interacting. At that time, the Fermilab accelerators could not make an intense enough neutrino beam for their experiment. However, the laboratory was in the process of building a new accelerator, and the feeder for this, known as the main injector, would be powerful enough to produce a useful beam of neutrinos at Soudan. By 1995, the new accelerator was starting construction, and a collaboration of physicists, including the original groups who had started the Soudan underground laboratory, was planning the MINOS (Main Injector Neutrino Oscillation Search) experiment.

A huge 5000 tonne detector was built in a new, bigger, cavern in the Soudan mine. This utilised yet another detection method. Charged particles passing through plastic, which had been loaded with small quantities of special chemicals, emit flashes of light (scintillate). These scintillations can be collected and delivered to phototubes which are similar in principle to those used to detect the Cerenkov light in the water detectors. By forming the plastic into narrow strips, sandwiched between plates of steel, the path of the charged particles through the detector can be followed, and by magnetising the steel plates, the curvature of the paths and thus the energy of the produced particles can be measured. From all this information, the details of the neutrino interaction, and in particular its energy, can be reconstructed. Then both the distance travelled (the 735 km from Fermilab) and the neutrino energy are known. A very similar (but smaller) detector was also built at Fermilab, so that by comparing the energy distribution of the neutrinos measured at Fermilab with that measured at Soudan, they could measure how any deficit depended on the energy of the neutrinos. If, as expected, this showed an oscillatory pattern, it would measure the difference in mass between the produced and oscillated neutrinos.

By early 2005, the accelerator, the neutrino beam and the two detectors were all complete, and the experiment started. Two to three neutrino interactions a week coming from the beam were recorded at Soudan, and after a year of running, a clear deficit of neutrinos was being observed. The experiment is still running, producing ever more accurate results. It is hoped that eventually it will show whether muon-neutrinos oscillate into electron-neutrinos and also tau-neutrinos, and also whether antineutrinos oscillate in the same way as neutrinos. There is even the possibility that fundamental differences between neutrinos and antineutrinos may be found. Such a discovery could shed light on the question of how our matter-dominated world emerged from the symmetric matter–antimatter universe that had been produced in the Big Bang.[xxvii]

Neutrinos 'Back to the Future'

Research into solar neutrinos had changed dramatically since Davis began. By 1990, the primary goal had become to understand the neutrinos themselves. When Davis began his quest, his team consisted of just a handful of scientists and engineers. A typical experiment today involves over 100 physicists in an international collaboration. Nearly all experiments are electronic rather than radiochemical. Whereas Davis found on average one neutrino a week, and knew nothing of its energy other than that it had to have enough to have been detected at all, the electronic experiments gather thousands of events each year, together with measurements of their energies and even the directions from which they have come. As we saw on page 115, it is even possible to show an image of the Sun shining in 'neutrino light'.

The most challenging frontier for solar neutrino research remains that of detecting the lowest energy neutrinos produced in the primary fusion reactions. Having such low energies, they have the smallest chance of interacting, but they comprise more than 99% of the total solar flux. The predictions from astrophysics about solar neutrinos are most precise for the lowest energies, below 1 MeV.

The SNO heavy water detector continued to study the Sun until 2006. At this point, the ten-year loan period from AECL expired. The remainder of the apparatus exists and, instead of heavy water, an organic liquid is going to be used, which emits flashes of light when charged particles pass through. These flashes will be brighter than was the case with heavy water. One consequence will be that SNO, in its new phase, will be able to detect neutrinos with lower energies than before. This may begin to teach us about the dominant primary production, from the basic proton fuel in the Sun.

There are also hopes that as more data accumulate, it will become possible to see if the fluxes change from day to night. At night, the neutrinos have to pass through the whole of the Earth, rather than just a couple of kilometres as by day. This predicted MSW effect in the Earth is small, so it will be a challenge.

The possibility of looking deep into space by means of the vast numbers of neutrinos that fill the void is an exciting goal. Astrophysicists believe that gamma ray bursts, which have puzzled astronomers, are accompanied by hordes of neutrinos. These are predicted to occur with energies greater than 100 trillion electron-volts, that is at least ten times larger than the energies that can be obtained for the primary beams in the most powerful accelerator on Earth, the LHC at CERN. To capture some of these neutrinos coming from the galaxy – and even beyond – huge underground detectors are involved.

Neutrino astronomy has moved out from laboratories in enclosed caverns and, to capture these most elusive particles, is now using natural features in the world as detectors. These new neutrino telescopes are underwater in the Mediterranean and Lake Baikal in Russia; they are under the ice in the Antarctic; they extend over a square kilometre, and have romantic names such as AMANDA and ICECUBE.

AMANDA is the Antarctic Muon And Neutrino Detector Array. It is buried under a kilometre of ice to detect high energy cosmic neutrinos coming from our own or other galaxies. In addition to solar neutrinos, there are neutrinos roaming the universe that are leftovers from the Big Bang, and also vast numbers pouring out from colossal stellar explosions.

There are indeed huge numbers out there, but they are relatively faint by the time they arrive here. To capture neutrinos from the Sun has required detectors with thousands of tonnes of material. Neutrinos from the far galaxy and beyond are likely to be relatively as faint compared to solar neutrinos as starlight is to daylight. To have any chance of capturing them requires detectors containing over a cubic kilometre of matter. It is obviously impossible to build such a thing in a laboratory, but the ingenious idea has been to use the ice in the Antarctic as a natural detector.

When neutrinos in cosmic rays hit atoms in the ice, muons can be produced. In turn these generate Cerenkov radiation, faint

149

flashes of blue light, as they pass through the ice. All that is required is to detect this.

Ice in the Antarctic is not like ice that we are used to on a cold winter's day at home. In the Antarctic, snow has fallen on ice for much longer than recorded history. At a kilometre below the present surface, the snow fell ten thousand years ago, soon after the last ice age. The pressure is so great that down there all the air bubbles have been squeezed out, leaving ice so pure that light flashes, produced by neutrinos, can travel hundreds of metres – undimmed. Photomultiplier tubes have been lowered into the ice, down shafts that are made by a special drill that sprays out hot water and melts a hole. The detector is attached to a long cable, lowered into the ice, which then freezes it into place. From then on it records data continuously. A lattice of these detectors awaits the tell-tale flashes of light, which signal a neutrino. The setup is so sensitive that it regularly records atmospheric neutrinos from all around the globe; some come from directly above the Antarctic, while others have travelled all the way through the Earth, from the North Pole.

Similar ideas are being developed in the northern hemisphere, but using water instead of ice. Since 1998, there has been a relatively small detector, with an area of a few thousand square metres, under the world's deepest freshwater lake, Lake Baikal in Siberia. Larger detectors than this, however, are needed to make truly sensitive measurements. A big array of phototubes is being built in deep natural trenches in the Mediterranean Sea. ANTARES is Astronomy with a Neutrino Telescope and Abyss environmental RESearch, a detector being built off the south coast of France near Toulon. Another detector, NESTOR, will be in the deepest parts of the Mediterranean to the south-west of the Peloponnese in Greece.

The galactic core of the Milky Way is completely obscured by dense gas and numerous bright objects. However, it is possible that neutrinos produced in the galactic core will be measurable by Earth-based neutrino telescopes in the next decade.

The aim is to know what there is in the universe that we cannot see in ordinary light and electromagnetic waves of any wavelength. The challenge will be to develop the right instruments to detect these neutrinos, measure their energies and identify where they have come from. If gamma ray bursts can be seen in neutrinos, we will be detecting neutrinos that have travelled across space for billions of years. As the ten minute travel time of solar neutrinos is vast on the scale of the nanoseconds in the laboratory, so are the journey times of cosmic neutrinos correspondingly greater again.

In travelling from the most distant parts of the universe, over such immense timescales, exotic properties of neutrinos might be revealed. It is possible that they will interact with the background radiation from the Big Bang. There may be surprises awaiting us that will turn out to be even more sensational than anything that has happened so far.

151

11

REPRISE

Eight decades after Pauli exclaimed that he had done a 'terrible thing', admitting that 'I have postulated a particle that cannot be detected', neutrino astronomy is at the threshold of enabling us to look into distant galaxies, and to find echoes of the Big Bang. A lot has happened since that seminal moment when Pauli invented the neutrino. The first stage in our story took 26 years, until the day in June 1956 when Clyde Cowan and Fred Reines decided that they were confident enough to tell Pauli that they had at last proved him right.

They sent him a telegram, which he received while attending a conference at CERN. He interrupted the meeting to read it to the audience. 'We are happy to inform you that we have definitely detected neutrinos. The [rate that we detect them] agrees with [what was] expected'. Pauli and his colleagues consumed a crate of champagne in celebration.[xxviii] Pauli paid up for the champagne that he had wagered years before, and also sent a grateful reply, thanking them for the news, adding the remark: 'Everything comes to him who knows how to wait'.

That remark turned out to encapsulate the entire history of neutrinos, and the various fortunes of the heroes in our story.

Reines' Encore

One leading physicist, Luis Alvarez, pithily commented that, after discovering the neutrino, 'What do you do for an encore?' If anyone expected a Nobel Prize for this discovery, they would have to wait. Reines set about an encore by continuing his quest 'to do the most difficult measurement possible'. The chance that a neutrino interacts with anything is small, and the theory implied that its interaction with an electron is the smallest of all. Reines spent 20 years attempting to measure this, making ever more precise experiments. He eventually succeeded, describing it four decades later in his Nobel Prize speech as the smallest cross section of any process ever measured.

However, he hadn't devoted 20 years just to this. His main activity from 1960 onwards had been to look for neutrinos produced naturally in cosmic rays. Collisions between cosmic rays and the atmosphere produce lots of pions, which decay into muons and neutrinos. It was such decays of pions produced in accelerators that had inspired Steinberger, Schwartz and Lederman in 1960. This intrigued Reines, who realised that cosmic pions must be producing showers of neutrinos. The challenge was how to detect them.

Having by chance paralleled Davis in the 1950s, when they were each looking for neutrinos at the Savannah River reactor, now their careers again followed a similar course. Davis went

153

a mile underground in South Dakota in search of neutrinos from the Sun, while Reines built an experiment two miles underground in a Johannesburg gold mine in pursuit of neutrinos from the cosmic showers. It was on 23 February 1965 that Reines detected the first 'natural' neutrino; up to that date the only neutrinos recorded had been made in reactors or accelerators.

Reines' underground experiments detecting cosmic neutrinos became his main interest. He was a leader of the IMB collaboration (page 112) when the first neutrinos from a supernova passed by in 1987. These data not only showed that our theories of a supernova are good, but also revealed things about the neutrinos. The results implied that there are just three varieties of neutrino: the electron, muon and tau type. In the 1990s experiments at CERN, completely unrelated to supernovae, also showed that three is the number.[xxix] This observation also agrees with theoretical cosmology, which can explain the relative abundances of the light atomic elements in the universe at large if three distinct varieties of neutrino emerged out of the Big Bang. So by 1995 three varieties of neutrino had been established, and the science of neutrino astronomy – covering supernova explosions and the Sun – had begun.

That year, 39 years after proving the existence of 'the most tiny quantity of reality ever imagined by a human being', and aged 77, Reines was awarded the Nobel Prize. The accolade recognised his lifetime's work with neutrinos: their discovery, determining their properties and inspiring the birth of neutrino astronomy. His acceptance speech[xxx] was fulsome in its praise for his one-time collaborator: 'Clyde Cowan was an equal partner. I regret that he did not live long enough to share in this honour with me.'

Koshiba in Japan

For Masatoshi Koshiba, the waiting was of a different kind. Born in Tokyo, in 1926, he had studied physics at the university there, graduating in 1951. He went to the USA to do his PhD, at the University of Rochester, spent three years as a research associate at the University of Chicago, 1955–58, and then returned to Japan. He spent his life in experimental nuclear and particle physics, gradually rising to be professor at the University of Tokyo in 1970. It was not until he was approaching 60 and retirement, that he came to the project that would make his name: inspired by Davis's attempts to detect solar neutrinos, Koshiba led the team that built the first full-scale solar neutrino observatory.

It had three different applications: one designed for, one that happened by chance and one that was completely unexpected. The first was its ability to take neutrinographs of the Sun, confirming Davis's results and, together with SNO, solving the solar neutrino problem. The second was utter serendipity when, just two months after they were ready, they heard that a supernova had been seen in the southern skies. The energy of each (anti) neutrino from a supernova is much higher than for a solar neutrino, which makes them easier to detect. As the detector was already sensitive to solar neutrinos, they checked their data and found the signal for neutrinos from the supernova easily.

Third, there was the unexpected bonus when they detected neutrinos from cosmic rays. As the different patterns of the Cerenkov radiation could distinguish electrons from muons, they were able to tell if they were detecting neutrinos of the electron or the muon variety. They had expected to see two of the muon

155

variety for every one of the electron type, but what they found was that there were fewer muon neutrinos than expected. They could distinguish neutrinos that had come from above Japan, on a journey of no more than 20 km, from those that had come horizontally, travelling about 1000 km, and those that came in from below, voyaging all the way through the Earth for 13,000 km. The further the neutrinos had come, the smaller was the fraction of the muon variety. So it was SuperK that could claim to be the first to have found clear evidence that neutrinos have mass.[44] This could explain why Davis had found fewer solar neutrinos than he and Bahcall had expected, and was consistent with another of Pontecorvo's ideas – that neutrinos can oscillate.

Among the many ironies in the neutrino story, the atmospheric neutrino saga and Koshiba's contribution are worth a comment.[45] The first experiments to have detected atmospheric neutrinos were in 1963, where these neutrinos were the unexpected by-products of measuring the penetration underground of cosmic rays (primarily muons) by Fred Reines and others. Despite the prediction of neutrino oscillations around this time by the Japanese team, and by Pontecorvo and Gribov, nobody was interested enough to put big detectors of neutrinos underground. Instead, all the effort was put into experiments using the much more intense beams of neutrinos that were beginning to be produced in accelerators,[46] but the 20 years of looking for neutrino oscillations at accelerators ended in failure.

[44] The 'disappearance' of the muon-neutrinos could have been either because they decayed, or because they changed form – oscillated. To find clear evidence of this would require the detection of what they had changed into. A loss of intensity could be because of either, but both possibilities required the muon-neutrino to have mass.

[45] Based on a letter from Don Perkins to the author 19 May 2009.

[46] See the story of Schwartz, Steinberger and Lederman in chapter 7.

156

Some scientists in those days did propose putting detectors several kilometres away from the accelerator, an idea taken up later (chapter 10), but the original plan at CERN was killed off by the CERN management. A proposal to put a detector on the far side of the Jura mountains, to detect neutrinos coming from CERN, was regarded as too sensitive at a time when CERN was trying to get support for their large electron–positron collider (LEP) a 27 km underground accelerator. The concern of the management seemed to be that if people realised that neutrinos could go through a mountain range, public opinion might start to worry about what they could do to humans. The fact that neutrinos do nothing to humans was an unwanted piece of extra 'education' that would be needed during sensitive negotiations. Politics won; the oscillation hunters at CERN lost.

By this time it was the 1980s and, as we saw in chapter 10, the failed attempts to detect decaying protons left atmospheric neutrinos as the only thing that the underground experiments could do. This was what led Koshiba to start proposing SuperK.

By 1990, five experiments in the USA, Europe and Japan were seeing the anomaly in the ratio of atmospheric muon- and electron-neutrinos. The irony was that it was not believed. In 1992, at the major international conference on particle physics held in Dallas, the reviewer of the field summarised why there were reasons to doubt the reality of the supposed anomaly.[xxxi]

The observation of atmospheric neutrino oscillations on Earth is in part due to the unique nature of our planet. The Earth's magnetic field determines the behaviour of the incident cosmic rays, and hence of the neutrino energies and intensities. The Earth's radius, which determines the distances that neutrinos have travelled from one side to the other, fortunately is just right to match the rate that neutrinos of these energies and masses oscillate. Furthermore, the Earth's density is small enough that only about one in a thousand neutrinos is lost while travelling diametrically through the core. This was all good luck.

Atmospheric neutrino oscillations were a fortunate by-product of a failed search for proton decay, and of the remarkable precision that large detectors filled with pure water could give. This was Koshiba's legacy – the inspiration behind SuperK – but he had retired by the time it produced the smoking gun. Koshiba shared the Nobel Prize with Davis in 2002 for their 'pioneering contributions to [neutrino] astrophysics'. At the age of 76, Koshiba was relatively a youngster.

Davis and Bahcall

Ray Davis, having survived to age 87, gratefully received a hugely justified Nobel Prize. He had tried to look into the Sun, and had devoted his entire career to achieve eventual success.

His long-time collaborator, John Bahcall, also had committed himself to a career-long quest. It was his paper and that of Davis, published back to back in *Physical Review Letters* in 1964, that had set the saga on its course. When Davis and Koshiba shared portions of the Nobel Prize for their experimental work, Bahcall was not included. There was speculation that had he and Davis written a single joint paper rather than individual ones, the outcome might have been different. However, this seems unlikely; the Nobel Committee quoted both Davis's paper and his own in their technical notes, so there was no possibility that Bahcall was in some way overlooked. The award recognised the primary and definitive acts of experiments that had created a new field of science: neutrino astronomy. Bahcall's calculations of the solar neutrino flux, but for which perhaps none of this would ever have begun, were indeed of singular importance in the story of 20th century science, but, at least in the opinion of the Nobel committee, qualitatively on a different plane.

Nonetheless, in the months leading up to the announcement, there had been widespread speculation that, now that everything had fallen neatly in place, this would be the year when the Nobel award recognised the neutrino chasers. Many physicists speculated that Bahcall's name would be in the short list of candidates, and were intensely surprised when it was not included. When asked for his reaction after the awards were announced, he generously said that they were richly deserved, and that he was pleased to have been 'mentioned in this distinguished company'.

He never will win the prize; in 2005, John Bahcall died, aged just seventy. He had survived long enough to see his life's work confirmed, and will be long remembered for his spontaneous remark when, in 2001, the SNO experiment announced its results, which proved him and Davis right: 'I feel like dancing, I'm so happy'.

Had neutrinos not oscillated, then it is likely that Davis would have measured the same number of SNUs that Bahcall had computed within a few years of starting the quest. The course of history might have turned out very different. That it took so long was because neutrinos carry identity cards, and can surreptitiously change them if given the right opportunity. Both of these facts were anticipated by Bruno Pontecorvo. Indeed, the whole course of neutrino physics had had Pontecorvo's theoretical stamp on it for more than half a century.

Bruno Pontecorvo

One of the joys in writing a book is that the plot does not always proceed as you expected. My original inspiration had been Ray Davis's singular dedicated quest for solar neutrinos, which culminated in his Nobel Prize. I had not anticipated that the identity of the central character in the plot would turn out to be Bruno Pontecorvo. It was Pontecorvo, no less than Clyde Cowan and John Bahcall, that I had in mind when I set the scene with: 'Longevity is an asset in the neutrino business. Not everyone would be so lucky.'

It was in 1934 that the young Bruno Pontecorvo had noticed that the radioactivity in Fermi's experiment behaved oddly. Fermi pursued this, and won the Nobel Prize. One of the long-term consequences of all of this was the development of nuclear reactors, which produced the intense source of neutrinos that Cowan and Reines used for their discovery.

Pontecorvo's entry into the neutrino story came in 1946, with his early discussion of the advantages of using chlorine as a neutrino detector. It was this idea that had started Davis on his life-long quest.

Pontecorvo's proposal that chlorine would be an ideal means of capturing neutrinos was correct; the problem was that reactors produce *anti*neutrinos for which this technique doesn't work. So the discovery of the neutrino by Cowan and Reines owed little to Pontecorvo. Had it been the case that reactors had produced neutrinos rather than antineutrinos, or that neutrino and antineutrino had behaved the same, then Davis would undoubtedly have made the discovery, and he and Pontecorvo probably would have shared the Nobel Prize for that. This was but the first possibility of a Nobel Prize that chance conspired to deny him.

Where Pontecorvo's idea bore fruit was in the quest for solar neutrinos. The Sun indeed produces neutrinos, not antineutrinos. Davis would be the first person to look inside a star, using Pontecorvo's idea to do so. However, it took nearly 30 years before people were convinced that he was right. Today, we know why it took so long: electron-neutrinos oscillate, whereby they had changed form en route from the Sun, and escaped Davis's trap.

Pontecorvo had even anticipated why solar neutrinos misbehave like this. He had first raised the question of whether electron-neutrinos and muon-neutrinos are different, shown how to answer the question in experiments, and when they were found to be different, even suggested that oscillations might be the reason for the shortfall in Davis's solar neutrino experiment.

Not only was Pontecorvo right here too, but the story of his insights is full of irony. Had there been no such thing as neutrino oscillations, Pontecorvo would have been right once (with his idea of chlorine as a detector) and solar neutrinos would have been seen by Davis at the expected rate. The irony here is that Pontecorvo was right more than once. The neutrino oscillations diluted Davis's signal to the extent that people doubted his results for nearly 30 years. Pontecorvo's suggestion that neutrino oscillations were responsible was largely ignored. It was not until 1998 that this began to be sorted out, and only in 2001 that it was finally settled.

The hypothesis of neutrino oscillations was a consequence of his earlier insight that neutrinos produced in association with electrons were in some way different from those produced with muons. Here he was a spectator as Lederman, Steinberger and Schwartz won the Nobel Prize for independently discovering this, and by the very means that he had suggested. He missed out here because, at a critical juncture in the neutrino story, he had chosen to live behind the Iron Curtain. His seminal papers appeared originally in Russian, unread in the West. The experimental facilities in the Soviet Union were not suitable for him to realise his dream, and the authorities refused to allow him to travel to the West, where he might have added a Nobel Prize to the prize he received from Moscow, the Stalin Prize.

Of all his ideas, the most far-reaching will surely be his insight, in 1959, that muon-neutrinos and electron-neutrinos are different. This led to the modern standard model of particle physics and the hypothesis that the eponymous electron-neutrinos and muon-neutrinos could swap identities by oscillating back and forth, so long as they had some mass. This idea was developed over several years into its mature form by 1967, a full year before Davis discovered the solar neutrino anomaly. It was while looking through an old edition of the journal, some years later, that I had stumbled on Pontecorvo's paper and 'discovered' that he was in Moscow.

Perhaps it is this paper above all that encapsulates the triumph and the tragedy of Pontecorvo's scientific career. It was because of neutrino oscillations that the Sun's neutrinos were diluted before arriving in the chlorine tank. Had they not, then Davis would have detected the full intensity in 1964 and been honoured at once, along with Pontecorvo: a second opportunity for them to have shared a Nobel Prize. Instead, the neutrino oscillations would at that time be a curse. Ray Davis had to spend 30 years trying to find why so many neutrinos from the Sun appeared to be missing. That would not be sorted out until the turn of the century, leading eventually to his Nobel Prize, aged 87, in 2002. Pontecorvo, however, had died in 1993, unaware of the great truths that he had expounded.

He never saw the phenomenon of oscillating neutrinos established, nor the way it is now being used to measure the subtlest properties of these ghostly entities. Today, these experiments promise to show how the material universe has evolved to its present form, with the real possibility that neutrinos hold the secret of why there is an excess of matter in the universe at large. Neutrino oscillations are established, and give hints that there is new physics awaiting discovery, but only if we do experiments at energies unseen in the universe since the Big Bang. The LHC at CERN will begin to expose some of these novel phenomena during the next decade. However, whatever the surprises awaiting us might be, none of the primary heroes of this tale will be around to build on the new visions.

Bruno Pontecorvo 'opened everyone's eyes with his original insights'[xxxi] Few scientists have produced such a wealth of far-reaching ideas without attaining a share of others' Nobel awards.

Sunset

This has been the story of the Sun no less than of the neutrino. Davis and Bahcall set out, intent on having neutrinos shed light on the Sun; instead the Sun has shed light on the neutrino. All through this tale what people set out to do has differed from what they would find.

In the 19th century, the question of how the Sun shines led quickly to conflict with geology and evolution. The greatest theorists of the time gave wrong answers all along – on the Sun's fuel, on the age of the Earth, on the implications for other areas of science. They were doing the best they could with what they then knew. With the benefit of hindsight, we can say that the Sun's heat and lifespan revealed that there is more in Heaven and Earth than were known in their 19th century philosophy.

The arrival of Einstein's theory of Special Relativity, and the precision measurements of the masses of hydrogen and helium, had nothing to do with this story – or so people would have initially thought. However, by the middle of the 20th century it had become clear that they were the crux of the plot.

It was these theories that enabled astrophysicists to work out how the stars burn. This showed that stars are nuclear fusion reactors in the sky, where exploding supernovae turn into neutron stars, and our nearest star, the Sun, is powered by hydrogen fusion. At first, this was all theoretical, based on remote observation, and on the results of experiments in the laboratories here on Earth. In the late 20th century, once the neutrino had been established, it became possible to detect neutrinos from the stars and learn directly what was going on inside them. Just a score of neutrinos from a supernova, spread over a few seconds with energies measured by flashes of light in an underground cavern, were enough to prove that a supernova is far hotter than the Sun, and that the result produces a dense neutron star. This is as the astrophysicists had predicted, and to me is perhaps the most remarkable synergy in all of pure science.

The Sun is near enough that neutrinos enabled us to make precision measurements on its deepest core. Here again the neutrinos showed that theory is correct, the actual energies of the neutrinos matched the predictions, as did their numbers to within a factor of two or three. Given that we might have found neutrinos hundreds of times more intense than expected, or even have found none at all, the fact that it was so near was remarkable. Today, when oscillations are taken into account, the agreement is very good. As had been the case with the measurements of helium and hydrogen masses decades earlier, here again the importance of precision measurement shone through.

The result of decades of research is that we are now certain that the nuclear reactions that give rise to neutrinos also make the Sun shine. This closes the wider scientific debate about the age of the Earth, and on the generation of solar energy, that began with Darwin and Lord Kelvin in the middle of the 19th century. The fact that the numbers were measured accurately enough revealed not only the workings of the Sun, but also that neutrinos are unexpectedly mysterious.

The exploratory phase lasted half a century; now neutrinos are becoming the means of making quantitative investigation of phenomena far away in the cosmos, and from deep in time. Not only are they looking into distant stars, but a new venture is bringing the story full circle. Radioactivity in the rocks is what helped to show that the Earth's age is billions of years, rather than millions, let alone thousands. That same radioactivity releases neutrinos. Now, by detecting these 'geoneutrinos', it will become possible to look deep into the core of our own planet.

Nothing of this was imagined when Becquerel discovered radioactivity, when Pauli invented the neutrino, or even when Reines and Cowan finally entrapped it. The long march to solve the solar neutrino mystery has created new branches of science: neutrino astronomy and neutrino geophysics. And the fact that neutrinos turn out not to be massless after all, is giving us clues to the realm of physical theories yet to be discovered.

The last word is with John Bahcall. This sums up the story of the science, and also could apply to himself and Davis, to Cowan and Reines, and, most poignantly, to Pontecorvo:

'If you can measure something accurately enough, you have a chance of discovering something important. The history of astronomy shows that it is very likely that what you discover will not be what you were looking for.' He then added, with typical modesty: 'It helps to be lucky'.

Notes

Chapter 1

 i Quoted in A Pais, *Inward Bound*, Oxford University Press, Oxford, England, 1986.

 ii That story is told in my book *Lucifer's Legacy*, Oxford University Press, Oxford, England, 2000.

Chapter 3

 iii H R Crane, *Reviews of Modern Physics*, 1948.

 iv F Reines, Nobel Address.

 v F Reines and C L Cowan Jr, *Physical Review*, vol. 90, p. 492 and vol. 92, p. 830 (1953).

Chapter 4

 vi A Prentice, *Mon. Not. Royal Astronomical Society*, vol. 163, p. 331, 1973; F Hoyle, *Astrophys J Letters*, vol. 197, p. L127; D Clayton et al. *Astrophys J* vol. 201, p. 489, 1975.

 vii *Nature*, vol. 284, p. 507.

 viii Quoted from D Wilson, *Rutherford – Simple Genius*, p. 206, Hodder and Stoughton, London, England, 1983.

Chapter 5

 ix Quoted in J N Bahcall and R Davis, 'The evolution of neutrino astronomy' in *Essays in Nuclear Astrophysics*, Cambridge Univ Press 1982, pp 243–285.

 x F Reines, *Ann. Rev. Nucl. Science*, vol. 10, p. 25.

 xi J Bahcall, *Physical Review*, vol. 126, p. 1143, 1962.

167

Chapter 6

xii Bahcall and Davis, 1982 op. cit.

xiii Quoted in Bahcall and Davis, 1982, op. cit.

xiv Bahcall and Davis, 1982, op. cit.

xv Bahcall and Davis, 1982, op. cit. p. 10.

xvi J Bahcall, 'Ray Davis: The scientist and the man', *Nuclear Physics*, vol. B (Proc. Suppl.) 48, pp 281–283, 1996.

xvii A Prentice, cited in Bahcall and Davis, 1982, op. cit.

Chapter 7

xviii B Pontecorvo, *Physical Review*, vol. 72, p. 246, 1947.

xix The full story is in *The God Particle*, L Lederman and D Teresi, Houghton 1993, pp 256–273.

xx M Schwartz, reminiscences at the Nobel Prize ceremony, 1988.

xxi G Danby et al. *Physical Review Letters*, vol. 9, p. 36, 1962.

xxii M Schwartz, Nobel Prize address, 1988.

Chapter 8

xxiii J Bahcall, *Physics Letters*, vol. 13, p. 332, 1964.

xxiv J Bahcall quoted in Nova on PBS. *Interview Dancing with Neutrinos* available at www.pbs.org/wgbh/nova/neutrino/dancing.html

xxv Don Perkins, letter to FEC 19 May 2009.

Chapter 9

xxvi Maki, Nakagawa and Sakata, *Prog. Theor. Phys.*, vol. 28, p. 870, 1962.

Chapter 10

xxvii See *Antimatter*, Oxford University Press, Oxford, England, 2009.

Chapter 11

xxviii As reported in F Reines, Nobel Prize speech, p. 251 in *Nobel Lectures 1991–1995*, ed. G Ekspong, World Scientific, Singapore, 1997.

xxix The details are in *Lucifer's Legacy*, op. cit.

xxx F Reines, Nobel Address http://nobelprize.org/nobel prizes/ physics/laureates/1995/reines-lecture.html

xxxi Hamish Robertson, *Proceedings of International Conference on High Energy Physics, Dallas, 1992*, American Institute of Physics, 1993.

xxxii Bahcall and Davis, 1982, op. cit.

Index